GOD, FAITH & THE NEW MILLENNIUM

Review coverage of Keith Ward's *God, Chance & Necessity*:

'At last, God is beginning to argue back.'

Clive Cookson in the *Financial Times*

'A witty, clear and probing critique of the atheist reductionists . . . This is a lively and important book . . . The debate between theism and materialism is of the highest significance and Keith Ward has made a worthy contribution to it.'

John Polkinghorne in the *Times Higher Educational Supplement*

'Beautifully and clearly written, accessible.'

Peter Atkins in *The Observer*

'What is fascinating and new about the book is Ward's unabashed animus and determination to take the fight to the scientists.'

Henry Porter in the *Guardian*

'A profound and richly satisfying book that illuminates better than anything I have come across the most important issue of our time and which sooner or later will consign atheism to the dustbin of philosophical ideas.'

James le Fanu in the *Catholic Herald*

'This is a book for which we have been waiting.'

Methodist Recorder

Other books by Keith Ward published by Oneworld

Concepts of God
God, Chance and Necessity
In Defence of the Soul

GOD, FAITH & THE NEW MILLENNIUM

Christian Belief in an Age of Science

KEITH WARD

ONEWORLD
OXFORD

GOD, FAITH AND THE NEW MILLENNIUM

Oneworld Publications
(Sales and Editorial)
185 Banbury Road
Oxford OX2 7AR
England

Oneworld Publications
(US Marketing Office)
PO Box 830, 21 Broadway
Rockport, MA 01966
USA

A CIP record for this book is available
from the British Library

ISBN 1–85168–155–8

Cover design by Peter Maguire
Printed in England by Clays Ltd, St Ives plc

Contents

Preface

When I came to teach at Oxford University, after many years of teaching philosophy in other British universities, I found myself almost immediately pitched into a rather heated debate about religion and science with two well-known scientists, Richard Dawkins and Peter Atkins. Professors Dawkins and Atkins regarded religion as some sort of intellectual disaster-area, and seemed to think that no intelligent and informed person could take religious beliefs seriously. Since I had been appointed Regius Professor of Divinity and was a Canon of Christ Church, I took some exception to their view, and over the next few years we had a series of public debates on the subject.

One result of this was that in 1996 I wrote a book, *God, Chance and Necessity*, in which I set out to refute their main arguments against theism and to argue that not only was belief in God compatible with modern science, but that the hypothesis of God was the best available explanation of an evolutionary worldview. The book attracted quite a lot of attention in the British press, and I received a huge number of encouraging letters, not least from some eminent scientists. But one of the things some of them said was: 'I can see the reasonableness of believing in God from what you say. But what has that rather abstract God got to do with actual religions like Christianity? Why go to Church or anything like that at all?'

This book is my attempt to show how Christianity fits into the worldview of modern science, and to show what intellectual belief in God as the cause of the universe has to do with things like going to Church and praying. It is about Christianity specifically, but a similar approach could well be taken by members of non-Christian religions, and I certainly do not see Christianity as the only religion that has an enlightened view on such matters. In fact, I think it is absolutely essential in the modern world for members of different religions and faiths to understand their own traditions more globally, and to work together to increase such understanding in every faith.

Different religions have different beliefs, and even different Christians have different beliefs from one another. I am not, in this book, trying to find a way in which such seemingly irreconcilable beliefs can be somehow harmonised. In fact, I think that such programmes will never succeed in producing some sort of universal agreement in religion, since orthodox believers in each religious tradition will reject them. Disagreement cannot be eliminated from religion, just as it cannot be eliminated from morality, philosophy and politics. It is important to ask how such disagreements arise, what might be done to prevent them being harmful, and how far they have got mixed up with cultural and historical factors that may be of secondary importance, as far as truth goes. But one has to do one thing at a time, and those are not my concerns in this book.

Nor am I trying to demonstrate that some form of Christianity is true, as though this was some sort of Christian propaganda. I want to look at what might be called a mainstream Christian view – a view which accepts, say, the Apostles' Creed as the basis of faith – and enquire how it could reasonably and non-hypocritically be interpreted, given a full acceptance of well-established scientific beliefs at the beginning of a new millennium. My conclusion will be that modern science does necessitate quite a bit of reinterpretation of traditional Christian ways of putting things. It would be surprising if that was not so, since our knowledge of the universe has increased amazingly even in the last few years. A result that may be surprising, however, is that the scientific worldview actually sets mainstream Christian beliefs in a context that seems to bring out richer depths of meaning which have always been implicit in them. And Christian beliefs provide a way in which the universe disclosed by science can be plausibly seen to have meaning and purpose. Contrary to what has sometimes been said, there is some sort of 'natural fit' between the scientific worldview and mainstream Christian beliefs, which does make Christian faith a plausible, though not provable, religious view in a scientific age.

Against that background, I have tried to show how the events surrounding the life of Jesus can be plausibly seen as playing a key role in the realisation of the divine purpose for this planet. They reflect the great cosmic themes of creation, divine love, and the uniting of all things in the cosmos to the divine life. Christianity has developed a sacred cosmology, expressed in highly symbolic form, which expresses these themes, and I have tried to show how scientific cosmology provides the factual, literal, background against which the spiritual vision of Christianity can be best seen.

So I have tried to present Christianity as a religion of truly cosmic scope, which can give insight into the meaning and purpose of this universe, the physical structure of which modern science has marvellously discovered. Such a view of Christian belief may seem new both to some Christians and to those generally interested in the nature and relationship of religion and science. It may show one way in which belief in the rather abstract 'God of the physicists' can be enriched by the insights of a particular religious tradition of prayer and worship. In this light, one might see not only that it is a very reasonable belief that there is a creator God, but also that it is natural and appropriate to worship and pray to God. Even though the claims of Christianity transcend what unaided human reason can establish – one does not become a Christian for scientific reasons – I hope to show that it is entirely reasonable to accept the Christian faith, and that if its claims are true it does provide a fuller understanding of the real character of this beautiful and awe-inspiring universe.

Introduction

God and the Scientific Worldview

As the world enters the third Christian millennium, many people feel that a new spiritual awareness is arising which has the power to transform and reinvigorate traditional religious views. Religious traditions carry many tried and tested insights into the reality of the sacred, and it would be irresponsible to ignore them. At the same time, all traditions have been called in question by two great global currents of thought. The rise of the natural sciences changes our view of the universe, and so of the context within which religious beliefs are to be interpreted. And an increased sensitivity to the spiritual resources of many different religious traditions urges a reassessment of the exclusivity and intolerance of some of our older views.

Christian beliefs were shaped in intellectual contexts very different from that of modern science, and at a time when few people had much idea of what would be required of a truly global perspective on religious beliefs. In the first millennium, the sophistication of Greek philosophy was used by Christians to construct a cosmic vision of the creation and redemption of the whole universe. But that tradition had not grasped the vast extent of the cosmos, the possibilities of historical change, and the revolution in our understanding of the universe that the experimental sciences would bring.

In the second millennium, Christianity, while it expanded throughout the world, underwent an internal fragmentation, and therefore a crisis of authority. The Orthodox and Catholic Churches split, and then the Reformation divided the Western Church still further. Towards the end of the millennium, the revealed texts were subject to searching analysis by the application of historical and literary criticism. Belief in miracles, in the soul, and in the existence of God were ruthlessly criticised. The patriarchal and hierarchical structure of the Church came under attack. Christians became aware that they were often seen by others as cultural imperialists and intolerant dogmatists, and many felt that the world had entered a post-Christian age.

As the third millennium begins, what will happen to Christian belief? Will it splinter into a myriad dogmatic sects, each certain of its own exclusive grasp of truth, all ignored by people of good sense? Will it fade away into an ultra-liberal confusion and puzzlement, full of goodwill, but without any clear beliefs at all? Or is there a possibility of integrating religious beliefs and scientific knowledge, commitment to a definite spiritual tradition and global awareness, loyalty to revelation and openness to new moral thinking? My own view is that the third millennium of Christian existence will bring a new integration of scientific and religious thought, the development of a more global spirituality, and a retrieval of some of the deepest spiritual insights of the Christian faith, which have often been underemphasised or overlooked. This book aims to present that view, to show one form that Christian belief might take in the third millennium.

The growth of the natural sciences since the sixteenth century was inspired by the thought that God's creation was meant to be understood by rational creatures, and that it therefore could be understood. The new understandings of the universe that science has brought have revolutionised all previous views of the universe. Even though science had most of its beginnings in a religious context, some scientists and philosophers argue that science somehow undermines all religious beliefs, and particularly belief in a creator God. Others – probably the great majority – agree that what science does is to give us an even greater sense of the wisdom and power of the creator. But some of them find it hard to connect the God who is a vast and almost incomprehensible cosmic intelligence with the God who seems to be the object of worship in churches, who is supposed to have acted on this small planet in the life of a young Jewish preacher, and who seems to choose such odd people to communicate with.

The most important thing about religion is not, of course, a speculative hypothesis about a cosmic creator, but its power to evoke some sort of experience that can give a sense of greater intensity and meaning to life. Religion is most basically about the fundamental problems of everyday living – how to cope with anxiety and hatred, how to achieve some sort of integration or happiness, and how to obtain some sense of meaning, purpose or value in one's own life. Religions take many forms, and they sometimes seem to breed anxiety and hatred, or they become indistinguishable from various nationalist or political movements. Religions do not necessarily offer a solution to personal or political problems: they can make them worse. Yet at the heart of most great religions, however much it may get corrupted by human weakness and perversity, is the attempt to find a way to overcome the greed, hatred and delusion that dogs human life. There is an attempt to find a way that leads one towards a better state of bliss, wisdom and compassion.

The Christian religion, like most human belief-systems, can sometimes seem to be nothing but a matter of intolerant dogmas or narrow-minded moralising. But at its heart the Christian faith tries to teach people to turn from self-centredness (picturesquely summed up in the phrase 'the world, the flesh and the devil'), towards a sort of experience which will free them from anxiety and hatred, and give a deeper sense of happiness and meaning. This experience Christians call the awareness of a being of supreme goodness, with the power to liberate them from self-centredness (this is what is called, in traditional terms, salvation from sin). Christians call this being God, and they claim to find God in three main ways – revealed in history in the person of Jesus, present within human lives in the form of the Holy Spirit, and transcendent in glory as the source and creator of all things.

This experience of God in threefold form (as a Trinity) is what makes Christianity distinctive as a religion. Just as the experience of being in love with another person can fill life with a sense of purpose and meaningfulness, so the experience of God can give courage, hope and joy to the whole of human life. God is not an object open to inspection by some scientific technique. Just as another person needs to disclose something about his or her inner thoughts and purposes if we are to understand that person, so God must disclose the divine mind if we are to know what God truly is. The Christian claim is that God has revealed the divine nature in and through the human person of Jesus of Nazareth. That is not the only place on earth where God is revealed, but for Christians Jesus is a distinctive and authentic self-disclosure of God. In Jesus the divine nature is shown to be a supreme reality of self-giving, sharing and unitive love, and through him that love is believed to be communicated to humanity. In such knowledge and love, Christians believe, human life finds its ultimate purpose and fulfilment.

The connection between the idea of a cosmic creator and the life of people in ordinary Christian churches lies in the fact that the God who can be experienced as a personal reality – the God of religion – is the same God who is the cosmic intelligence, the creator of the universe. Of course humans can never experience the cosmic intelligence as it is in its fullness. But there is no reason why such an intelligence should not make itself known in a personal way, if it so wills. Christians believe that the creator does intend to make itself known to finite creatures, and that, in fact, one main reason why there is a universe at all is to create beings that can find happiness in knowing and co-operating with the creator. There is a link between a general cosmic purpose in creation, and particular historical events on this planet that are claimed to reveal God and save human beings from self-centredness. The link is that one part of the cosmic purpose is to create beings who can find happiness by knowing and loving God. If a being comes to know and love God, that must happen at a particular place and time, and it must happen in response to some self-disclosure of a God who makes the divine known. So there must be particular times in history when God is revealed, and at those times the purpose of the creator comes much nearer to fulfilment. Strange as it may seem, if there is a purpose in creation, it is very likely that one important part of it will lie in particular events on this planet which reveal God's nature and purpose.

Naturally, this will only seem plausible if one thinks there is a purpose in creation. That needs to be investigated further. But if there is a vast intelligence behind the universe, it is reasonable to think that it has brought the universe into being for some purpose. It will then be natural to try to find some evidence of what that purpose is. And that is where the religions of the world, Christianity among them, try to point to particular key experiences which, if they are genuine, do give evidence of such a purpose.

The idea of God is not primarily dreamed up as a hypothesis to explain why the universe is the way it is. It is not part of a scientific attempt to explain the basic nature of the physical universe. Nevertheless, if God is the creator of this universe, it is only to be expected that discoveries the sciences make about the universe will give a deeper understanding of the nature of the universe's creator. Further, it is to be expected that the concept of God will provide the best explanation of why the universe is the way it is, and a much better explanation than hypotheses that deny the existence of God. The modern scientific worldview and the Christian revelation can, and do, interact fruitfully to provide a coherent and illuminating picture of human existence in this extraordinary universe. I want to show that this is the case, and explain why some scientists deny the existence of God, partly by misunderstanding the way in which God explains the universe, and partly by misunderstanding the limits of scientific theorising.

The real reason people believe or disbelieve in God is not to do with science, but with highly personal factors that predispose people to be either sympathetic or antagonistic to the experiential and moral claims of religion. If one has had experiences in religious contexts which have been positive and life-enhancing, which have helped one to overcome hatred and greed and achieve a more integrated and committed life, one will be well disposed to the claims of religion. If one has suffered from censorious, petty-minded or intolerant religious believers, one will naturally be much less sympathetic. It may be that one has simply not had any experiences that seem to be of a transcendent or spiritual reality. Or perhaps personal tragedies have made one sceptical about there being any moral order in the universe at all. There are many different reasons for being religious or non-religious, but usually personal experience, not abstract speculation, is the decisive factor.

Speculation, the attempt to construct a coherent and plausible account of the nature of the universe, is still obviously relevant to the acceptability of religious belief. In our day, it is science that gives such a general speculative, but well-evidenced, account. It is sometimes argued that science is antagonistic to religious faith. I aim to show that this is simply not true, and that an informed scientific view of the universe is not only compatible with Christian belief, but actually lends quite strong support the claims of some forms of Christianity. Christian belief in the third millennium can and should be scientifically informed, globally aware, a source of distinctive insights into the human condition, and a great enrichment of human life and welfare. Whether or not it will be is up to Christians themselves.

1

Christianity and the Scientific Worldview

Traditional Religious and Modern Scientific Pictures of the Universe

The fundamental idea of God that is found in Jewish, Christian and Muslim traditions is that God is the one and only creator of everything other than God. God is 'the Father' of the universe, a traditional metaphorical expression for the one who brings the universe into existence and sustains its life. This idea of God might be accepted simply because it is revealed in the Bible and the Qur'an. But it is an idea that arises naturally as one reflects on the nature of the universe in which humans exist. In fact, modern scientific knowledge of the universe suggests the idea of a creator with almost compelling force, yet it can look as though the scientific idea of the universe is in conflict with the traditional religious idea.

This is because, when the ancient Scriptures were written, the universe was believed to be relatively small, both in space and in time. In the first chapters of Genesis, the earth was the centre of the universe, a disc floating on water, covered by the hemisphere of the sky, from which the stars, sun and moon were hung as 'lamps' to measure the seasons, days and nights. It had not existed for very long – it had been created in six days, about four thousand years before Christ. And it would probably come to an end quite soon, in a few generations at most, when the sky would fold up, the star-lamps would fall to earth, and the whole created universe would be destroyed.

The modern scientific view of the physical universe is very different. The earth is a small planet in a sun system at the edge of just one galaxy of stars out of many millions of galaxies. It is dwarfed into physical insignificance by a huge and expanding universe, which has no centre but is spread out like the surface of an expanding balloon. The planet earth has existed for billions of years, and it will continue to exist for perhaps another five billion, when it will become uninhabitable as the sun turns into a red giant star and incinerates all life on earth, before collapsing into a tiny white dwarf star. Human life evolved from simpler organic forms, and has existed for about two million years, about ten thousand of them carrying remnants of fairly advanced cultures. The planet earth may indeed cease to exist at any moment, through some cosmic catastrophe. But if it does, that will not be the end of the universe. It will be an event of hardly any significance to the universe as a whole, which will go on existing for billions of years, perhaps evolving many forms of life more advanced than the human, for all we know.

Thus the picture of the universe given by the ancient religious traditions is very different from the picture given by modern science. In the traditional picture, human existence seemed to be the most important thing in creation. Heaven was populated almost entirely with humans, and God's purposes for humans were assumed to be coincident with God's purposes for the whole creation. Modern science gives us a picture that looks very different. Human life seems very small in relation to the whole universe. We seem to have lost our central place in creation.

Is Human Life Insignificant?

To some people, it seems that the scientific view deprives human life of significance altogether. The best available view in physics is that the universe originated about ten to fifteen billion years ago, in a state of infinitely compressed gravity and energy. It exploded into an expanding balloon of space–time, and as it began to cool, the basic forces of electricity, magnetism, gravity and the strong and weak nuclear forces were created. Basic elementary particles formed and began to join up into simple atomic structures. The simplest atoms of hydrogen and helium formed swirling clouds of gas, which condensed into galaxies, stars and planets. Stars grew and collapsed, in their death throes creating more complex elements, such as carbon. At a later stage still, planetary systems were seeded with carbon atoms, and on the surface of at least this planet, earth, the extremely complex molecular structures of RNA and DNA gave rise to the self-replicating forms that we call 'life'. This, in turn produced even more complex structures with consciousness and some capacity for action.

The fifteen-billion-year history of the expanding universe has been one which has produced amazingly complex structures on one tiny planet, and the process is still continuing. We do know, however, that the universe has a finite life span. If it goes on expanding, the law of entropy will ensure that all complexity and organised structure is eliminated. If the universe contracts again, under the force of gravity, it will end in a Big Crunch, in which all things will be destroyed. Either way, the complexity of human life, which was always a freak event in the history of the universe, will inevitably be destroyed. The universe will simply run down, and in it the whole story of humanity will have been a peripheral sideshow, the flicker of a dragonfly's wing, present only for an instant in the long, blind, purposeless interplay of physical energies that is the real nature of this universe.

That is how it looks to some people. And yet we must be very careful to separate well-established scientific theories about the history of the universe from the significance we give to the process. It is quite true that in the scientific account, human life is a tiny flicker in the vast spatial distances and temporal aeons of this universe. We can look out on a clear night, see stars millions of light years away and feel reduced to insignificance by the vastness of nature. But we should bear in mind that size is not to be confused with importance. I might be a tiny speck in comparison with the starry sky. But if, on all the stars in that sky, there are nothing but lifeless, unconscious nuclear reactions and vast swathes of gas and dust, my little life may be more important than all of them put together.

Just suppose that there is a God who created the universe. No doubt the creator enjoys the beauty and variety of the physical universe, and the intergalactic clouds and the galaxies themselves are important to the creator because they express his power and wisdom: 'the heavens declare the glory of God' (Psalm 19: 1). But if it is the creator's wish to produce conscious agents who can come to understand the universe and share in God's appreciation of it, to know the creator of the universe, and enter into a relationship of loving co-operation with the creator, then the goal of the whole cosmic process will be the existence of such conscious agents. In that sense, however many billions of galaxies there are, they will all exist as parts of a process whose goal is the existence of human beings. To be exact, the goal will be the existence of rational conscious agents, of which there may be many in the universe. But we only know of one case so far, and it may be the only case. Either way, human beings are one of the forms of life that are the purpose of creation. It is quite consistent with modern scientific knowledge, after all, to think that human beings are more important than the stars and galaxies. Humans may well be the reason why stars and galaxies exist, and they are at any rate one of the reasons why they do.

It may seem a little arrogant to think that the whole of space and time, billions of years old and billions of light years in extent, exists just to produce human beings. In fact, recent research in physics shows that it is not a ludicrous thought. It has become increasingly clear in recent years that the basic constants and laws of physics need to be very fine-tuned to generate the sorts of complex order that exist in our universe. The four basic forces that make up our universe – electromagnetic, gravitational, and the weak and strong nuclear forces – establish a set of relationships between the fundamental physical particles (at present thought to be quarks) that enable them to build into the protons, neutrons and electrons that form atoms of the various elements. Those atoms in turn combine in various ways to form molecules, and they build the extremely complex physical structures that enable living organisms to develop. Such organisms develop central nervous systems, and eventually brains, which can mirror and react to their environments. By a further development of the cortex in some of these brains, truly self-conscious agents come into existence.

All this incremental structural complexity happens in accordance with the fundamental laws of physics, which were built into the universe at the moment of its origin, the so-called 'Big Bang'. Some physicists have calculated that the time it would take for complex self-conscious agents to evolve from simple atomic structures, in accordance with the basic laws of physics, would be about fifteen billion years. During that time, the visible universe has been expanding at the speed of light, so that the size of the universe should now be fifteen billion light years across – which it is. In other words, the universe needs to be just about the vast age and size it is, if human beings are to evolve in accordance with the basic laws of physics. If the creator wanted human beings to evolve through the successive application to elementary physical particles of a basic set of elegant and simple physical principles, then a universe of just this size and character would be the one to choose.

It turns out that human beings are not peripheral to the universe after all, even if they are dwarfed in size by the physical cosmos. They could well be the reason why the whole cosmos was created. The moral of the story is that size is not everything. A mind that can understand this universe, and even begin to change it, is in many ways of more significance than a billion

billion light years of unconscious interstellar dust. The generation of minds that can come to know and love their creator may in the end be what gives meaning and significance to the whole cosmic process.

The Ending of Life and the Meaning of Life

But what about the inevitable destruction of the universe? Does that deprive human life of significance? The thought that it does rests on another misconception. Just as it is mistaken to think that the vast size of the universe reduces humans to insignificance, so it is mistaken to think that the fact that all sentient life will come to an end deprives such life of significance. A life does not have to last for ever in order to have meaning and value. It has meaning for as long as it lasts, and the thought that it will end may even give it greater meaning. Think about listening to a Mozart symphony. For many people, that will be an experience of great significance. It will be an experience worth having, just for its own sake. Let us call that an experience of intrinsic value. It is intrinsic because we do not value it just as a means to something else, like having a drink in the interval, or getting home after the concert. It is a value, because it is something we would choose to do for its own sake; it is something worthy of choice by any rational and sentient being (whether or not they actually choose it is another matter; the point is, it is worth choosing).

What gives life significance are experiences of intrinsic value. There are millions of different sorts of experiences like that, from playing skittles to enjoying a good meal or walking in the mountains. Listening to a Mozart symphony is just one of them. Now Mozart symphonies all come to an end. Does that mean they are not significant? Not at all. It is safe to say that they would be less significant, or at least less enjoyable, if they went on for ever (people who do not like music may dread the very idea). When we know the symphony will not last for long, we savour its sounds with even greater intensity, attention and enjoyment. Here is something that we may never hear again. This is our only chance to get an experience of very great value. We must grasp it with both hands: *carpe diem*, seize the day!

So the knowledge that a Mozart symphony will end gives it more, not less, significance. Moreover, its significance does not lie in its ending, in its last chord. That last chord, on its own, has very little significance. Even the final phrase, however beautiful it may be, could hardly be regarded as the reason why the whole symphony exists. The symphony does not exist just to produce that last phrase or two. What matters is the whole symphony, the way the themes come back in varied forms, the way the different movements contrast and build on one another, the way the music forms a pattern and a whole which has a sort of overall completeness and integrated beauty of form.

So we should not look for the meaning of a symphony just in its final few bars. We hope that they will form a fitting resolution of the work, but without the whole preceding set of sounds, they would mean much less. The meaning of the symphony is apprehended only by apprehending the whole work, with each part in its proper place. That is why it is not really satisfactory only to hear the 'purple passages' from longer pieces of music, without appreciating the context into which they fit and which gives them a rather different experienced quality. You have to hear the whole thing; and if it is by a gifted composer like Mozart, that will introduce you to forms of beauty you would never otherwise have known.

In a similar way, the significance of human life does not lie in its closing moments, whether we are talking about the cosmic death of the human race or the deaths of its individual members. The significance lies in the process of the life itself, and the way in which its various elements fit into an overall pattern within it. When people complain that life is meaningless, they often mean either that they can discern no intrinsic values in it, or that they cannot see how the events that happen to them fit into any overall pattern at all. To see the meaning of a human life would be to see the distinctive values it realises, which would otherwise not exist at all, and to see how its various elements fit into a unique, complex and integrated pattern. It is not to discover that there is some last state it arrives at which is somehow the purpose of the whole life. In a word, the purpose is not the ending, but the process. A life has a purpose if it has a pattern and a set of realised values that are unique and distinctive to it.

So the universe may have a purpose, if it is a process in which distinctive intrinsic values are realised in a complex, unique and integrated pattern, even if the last temporal state of that universe is one of dark, lifeless emptiness, or the incandescent flare of the Big Crunch. It makes sense to see human beings, or perhaps rational beings in general, who do realise complex and unique intrinsic values, as the purpose for the existence of this universe, even if they all die out long before the universe comes to an end, as it inevitably will.

Why Cosmic Evolution?

This helps to answer the question of why God has created rational beings who evolve over billions of years, rather than just creating them fully fledged and instantaneously. Humans may be the goal, or part of the goal, of the physical process. Nevertheless, what is significant about this universe is the whole process, and not just the attainment of its goal. Human beings recapitulate in themselves the history of cosmic evolution. They carry in their physical structure, in their genes and molecules, the history of their past. They are parts of that long physical process that is becoming aware of its own nature, and is beginning to direct it responsibly. The goal is not just the culmination of the process, but the whole process itself, which is one of generating from within a material cosmos the ability of that cosmos to understand and take control of its own being.

Human beings are parts of the cosmic process in which one can see this understanding and control begin to exist.

They are not mature, conscious agents without a history, who just begin to exist, with the whole material cosmos as an unnecessary backdrop to their activity. They are highly complex material structures, carrying within themselves the history of their development, integral parts of a wider material order. It is that order which they can understand and help to control, and human life should be seen as just one part of a wider process of creative development.

The process is one of creative emergence, as new forms of intelligibility and beauty successively come into existence out of earlier and simpler structures. One can see this whole process as if it were a work of creative imagination, continually building complex structures out of simpler elements, and always moving towards the capacity of the process to know and shape itself. If it is such a work, the whole process will have a beauty, intelligibility and imaginative flair that will be of great value to the creative intelligence that creates it and continues to move it towards that final goal which has been implicit in it from the first. Humans are not spiritual substances dropped into the material world as alien intruders. They are parts of a continuum of growing complexity in the material order, realising possibilities implicit in that order from the beginning.

For most of its existence, it may seem that all the beauty and intelligibility of the universe will be unappreciated, because there will be no one there to appreciate it. But if God is the imaginative creator, the cosmic artist, then of course God will know and appreciate the whole cosmic process. God will be the active power that urges the universe to realise new actualities out of the indefinite array of possibilities that are implicit in its origin. The whole process will be of value, because it is valued and appreciated by God. Not only that, it is in itself an expression of the imaginative and creative activity of God. Christians should never think that only human beings are valuable, or even that only the planet earth is valuable. The whole universe is valuable, to be appreciated and cared for as much as it is within our power, because it is shaped by a creator who cares for beauty and truthful understanding.

The process of creative emergence is one that generates from itself communities of beings who can know and shape its future. The physical universe is not just like a piece of clay in the hands of a potter. It is rather as if the potter could make clay figures that come to life, feel, think and act. They even begin to understand their own nature and shape it themselves. Such beings might come to appreciate their world and shape it in co-operation with their creator. So, in our universe, some animals might begin to share in the understanding and imaginative creativity of God. A Christian would say that they learn to co-operate with the divine creative Spirit in the task of making the material world a sacrament of the divine Spirit, a world in which wisdom, creativity and friendship can be celebrated and enjoyed. Such a sharing of understanding and creativity is not unlike the sort of understanding and shared activity between human beings that we call love. That is why Christians are able to say that the universe is created because of the love of God – the desire of God that there should be finite beings that can share in their appropriate way in the understanding and activity of God.

Physicists do not typically speak of such things as the love of God. They might even think that the cosmic intelligence seems to show a sort of cold indifference to human life. It often seems to them that talking about 'love' makes God much too anthropomorphic, much too small, to be the creator of this vast universe. But even speaking of a cosmic intelligence entails seeing the creator as conscious and wise. It is not really too large a step to see the creator as intending to create beings that can have their own wisdom and creative power, and can then move on to share the wisdom and power of God. Such a God can be spoken of as wanting creatures to know and love God, to have a relationship with God which might properly be described as a loving relationship.

If the creator designs the universe to produce beings that find their fulfilment in a loving relationship with the creator, we might well say that the universe is created by God in order to realise love. Of course, before we could say that, there would have to be some creatures that do seem to experience such a loving relationship with God. That is just where Christian faith comes in with the claim that a loving relation to God can be experienced, and has been realised in history, at the very least in the person of Jesus.

The personal experience of religion adds to the intellectual perception that there is a creator, and makes it possible to think of that cosmic intelligence as an intelligence that designs the universe to realise many forms of loving relationship. The creator may, by a sort of limitation of its infinity, enter into such loving relationship with creatures. That would not make God too small or too human. It would make God big and powerful enough to enter into personal relationships with creatures, while losing nothing of God's own infinite wisdom and power. It would add a personal dimension to the unlimited reality of God.

The scientific view of the universe does lead one almost inevitably to think of the wisdom and power of an unimaginably awesome creator. Christian belief does not take away from that at all, but totally accepts it. It then adds, on the basis of personal experience, that the creator wills to limit and express the divine nature in such a way that it can establish a personal relation, not unlike that of human love, with some of its creatures. In doing so, God realises one of the great purposes of the existence of this universe, the development from a material cosmos of beings capable of finding happiness and fulfilment by knowing and loving the creator.

One might envisage three stages of a created cosmic process which has this purpose. First is the stage of imaginative creativity, in which the creator shapes ever more complex emergent forms out of the primordial energy, the 'great deep' of potentiality. The Genesis picture of the creative Spirit moving over the waters of the great deep, the sea of primal chaos, expresses this stage exactly (Genesis 1: 2).

Next is the stage at which consciousness emerges as a property of the cosmos, and begins to co-operate with God in shaping the cosmos to a form in which it can express the divine Spirit itself. Again, the Genesis picture of humans being given dominion over the earth, with the responsibility of making it fruitful, is a very apt one (Genesis 1: 28).

Third is the consummation of the cosmos, in which it becomes an expression of perfect beauty, without conflict or defect, fully expressive of the divine nature as love, shaped by communities of love which share in the expression of the divine love. The New Testament picture of a universe united in Christ depicts such a fulfilled universe (Ephesians 1: 10).

In this way one can, from a Christian viewpoint, see why a creator God might originate an emergent, evolutionary cosmic process. It would be a process culminating in the goal of a sacramental community of love, but the process leading to that goal would be an important factor in the value of the goal itself. In that final consummation, the whole process could be summed up and seen as having from the first been directed towards a partly self-shaped conscious union with the creator.

2

The Trinity and Creation

God and the Christ

Far from the biblical and scientific pictures of creation being at odds with one another, when one looks more deeply, a surprising consonance of Christian revelation with the theory of cosmic evolution is apparent. Biblical revelation leads us to see God as the creator of all things. It leads us to see the cosmos as created through Wisdom, which delights in the creative process (Proverbs 8). It sees the cosmos as developing from primal chaos to ordered creation (Genesis 1: 2), through the creative activity of the Spirit, patterned on the archetypal Word of God. Human beings are not seen as disembodied souls inserted into alien matter, but as formed of matter (dust), which is shaped to become a finite image of God, to be self-aware and self-directing (Genesis 2: 7).

The doctrine of the Trinity, which is sometimes made to seem so obscure and difficult, can be a great help in understanding the nature of creation. The Christian vision of the creator God sees in the act of creation itself a threefoldness which expresses the nature of God as Trinity. The first element of this threefoldness is the unoriginated source of all being, beyond name and form. This source always remains transcendent to every finite reality. It is called 'the Father' metaphorically, to designate it as the primal cause of all beings. It would be wrong to think that this primal origin is either male or female, or limited in any way to finite forms of being. The term

‘Father’ originated at a time when it was thought that the male was the source of the vital principle of life. In many human cultures, there is a separation of female and male principles, *yin* and *yang*. In such separation, the male symbolises power and strength, domination and protection. The female symbolises beauty and tenderness, care and consolation. These topics have become very controversial in modern discussions of sexuality, but there is little doubt that they are traditional characterisations, and they remain powerful symbols. There is also little doubt that God must possess all such properties in their perfect form. There is, as Jung argued, good reason to include a ‘feminine dimension’ in the being of God. God relates to creation with ‘motherly’ care and compassion. Yet as the one and only source of all, the primal cause is named ‘Father’, to symbolise the indestructible power of being, which conquers chaos like a warrior and has the strength to ensure that its purposes will ultimately be achieved.

Jesus called God ‘Father’, *abba*, a term which has a much more intimately personal connotation and qualifies the impression of brute power that might otherwise be given. One should perhaps often seek other terms for the creator – including ‘Mother’, ‘Lord’ and ‘lover’. Yet the term ‘Father’ is retained as a primary symbol for the creator by Christians because Jesus used it, and to signify that unsurpassable power, even if it is power exercised always in love, is the sole prerogative of God the creator.

The second element is the self-expression of that One in a particular name and form. In Christian thinking, this first self-expression of the Father is called the Logos or the Word of God. Through the Word of God all things are created. As a thought takes form in the mind, so the Word or Wisdom of God takes form in the depth of the divine being. As a child takes substance from the parent, so the expressed Word takes substance from the Infinite Depth of Being. For this reason the Word is also called the Son of the Father, an ‘only-begotten’ son, because he is the unique expression of the depth of being.

Just as thought gives utterance and manifests the nature of mind in finite name and form, so the Son manifests infinity as its express image, its expressed name and form.

This does not mean that there is only one finite form of the divine – say, a human form. For that eternal image which is the Word is not limited to one particular form. It is so rich and comprehensive that it includes every finite form. In this image the whole universe is formed. There may even be other universes, other forms of space and time, and they too will be formed on the endlessly rich archetype of the eternal Word. He is the pattern and archetype of creation. Colossians 1: 15 is referring to this divine Word when it says, 'He is the first-born of all creation; for in him were all things created . . . through him and for him.' The Word is not a man, but the first-born, the visible image of the invisible infinity of God. Through him, the exploding energy of this primordial universe takes shape. He is the matrix of creation, the mould in which it is shaped and its primal origin.

'In him all things hold together' (Colossians 1: 15), and God's purpose is 'to unite all things in him, things in heaven and things on earth' (Ephesians 1: 10). In him, the manifold beings brought forth by the emergent universe exist and take form in their intelligible order. He is also the purpose of creation. By him, all finite beings are drawn into unity and to their proper perfection within the whole. The divine Word and Wisdom is the pattern of the beginning and end of this cosmos, which exists in order to bring to realisation the purpose framed in the mind of God at its beginning.

Creatures are to shape themselves upon the divine Word, becoming manifold forms of its life in time. In so far as things are true to their divinely intended natures, they express the divine life. They become parts of the 'body' of God, expressing the divine purpose in action, temporal images of eternity.

When creatures are called parts of the body of God, what is meant is that finite persons are to be related to God in somewhat the same way as human hands and mouths are related to the human mind. Human mouths give utterance to the thoughts of the mind. Human hands give expression to the purposes of the mind. The human body is the means by which the mind shapes the world to beauty and realises its potential for good. So the cosmos, directed by God, gives rise to sentient, creative creatures that can bring all its potentialities for good to fulfilment. In this sense, creatures are to be members of 'the body of Christ' (1 Corinthians 8: 6), and therefore of the body of God the Word, uniting all things in harmonious order as true expressions of the divine life.

The word 'Christ' means 'the anointed one'. It translates the Hebrew word 'Messiah', which refers to a being destined to appear as the divinely appointed saviour and king of Israel. Just as the kings of ancient Israel were anointed to symbolise their authority to rule so this being is given divine authority to rule not only Israel, but the whole creation. In Paul's writings, the word 'Christ' is sometimes used to refer to the eternal Word, who delivers and rules creation by uniting it in himself to God. It was this eternal Word who, according to Christians, united the human person of Jesus to himself, so that Jesus became 'the Christ' for this planet, the one who would deliver and rule in the name of God. The eternal Christ, the Word of God, is, however, not to be identified solely with the human Jesus – even though Jesus is the Christ for earthly history – because the Christ is the deliverer and ruler of the whole cosmos.

The deliverance of the cosmos, when all things are united in harmonious order, as yet lies in the future. 'The creation waits with eager longing for the revealing of the sons of God' (Romans 8: 19), for the manifestation of selves that can truly become members of the body of the eternal Word, vehicles of the divine life in the created cosmos. In the present, humans are called to work towards this goal, the 'kingdom of God', which Jesus reveals.

This is the Christian vision of the purpose of this cosmos. It sees this purpose as somehow having been impeded or frustrated, and yet as having been clearly foreshadowed in the life of a particular man, Jesus, son of Mary, and to a lesser extent in the lives of those who have been called to follow him. The Christian faith is above all a hope for the setting free of creation from its bondage, to allow the divine purpose to come to full fruition. The story of that bondage, and of how it came to be, is part of the Christian vision, too. But what is central is that creation is for a purpose; that the purpose will be realised, by divine power; and that there is a living Self of the universe, its pattern and goal, which is the eternal Word, the essential self-determination of the one creator God.

The Spirit of God in Creation

There is a third essential element in the creative activity of God. From the unknowable source of being issues not only the Word that gives form, but also the dynamic energy of the Spirit that gives life. This is the third element of the threefold Godhead. The Word is the finite image of infinity, embodied in the physical universe through the action of the Spirit, which prepares the way, realises the pattern, guides to the goal, and so expresses the eternal Word creatively in time. The breath of the Spirit stirs the deep ocean of potency at the beginning of the creation of his space–time. It drives the primordial energy of the universe towards life and creativity. It stirs restlessly through the elements of physical being, urging them towards complexity and order. It urges the stars to grow and die, to spawn the planets and to seed them with carbon and oxygen, to compact the dark immensities of space into the precarious forms of life. Over aeons of time it moves and prompts the primordial energies until one tiny part of the material order comes, for the first moment in that immensity of time, to realise that it has being. Consciousness and feeling come into being within emergent matter. The created order begins to know itself. This small blue planet is one such place, perhaps not the first, perhaps one of countless worlds of stirring consciousness. Whether on one world or on many, it is the Spirit, in her dynamic driving power, that shapes the first faint images of the Self from the dust of dying stars.

This is the threefold face of God: the measureless source of all; its primal self-determination, the Supreme Self, pattern and goal of creation; and the dynamic Spirit of Life, embodying that pattern and realising that goal in the temporal universe. It seems to me that it is the Spirit that can be best represented as the 'feminine dimension' in God. I say this with some diffidence, since I realise that it may be seen as a lame attempt by a male theologian to make a nod in the direction of feminism. Yet there are some good reasons for regarding the Spirit as feminine. In the passage from Proverbs 8 already referred to (page 31), the Hebrew word for 'wisdom' is *Chokmah*, a feminine noun, and in the whole passage Lady Wisdom is portrayed as playing before the face of God in the first creation. Wisdom, in that passage, has sometimes in Christian tradition been taken to refer to Christ, and sometimes to the Virgin Mary. The latter attribution is plainly unsuitable, since Mary is undoubtedly a particular human being who came into existence at a much later point in history. The former is possible, but it seems more suitable to refer *Chokmah* to that Spirit or 'breath' – *Ruach* – that blew over the face of the deep at the beginning of this creation. More generally, wisdom is regarded as female in many ancient religions: in the Indian traditions *shakti* is the female creative power that shapes the whole of the cosmos, giving it life and vitality. One may think that the Indian traditions have been more successful than the Abrahamic in incorporating feminine aspects into the divinity. The resources for doing this already lie in the biblical traditions, though they have not usually been exploited.

One reason why the Spirit has been referred to as 'he' in Christian tradition is that the Spirit is regarded as the Spirit of Christ, and Jesus the Christ was of course a male human being. Yet I have emphasised that the eternal Christ, the liberator and ruler of the created order, is neither human nor male. It might preserve this thought better to regard the Spirit which makes

Christ known, which is said to have filled the life of Jesus, and which brings the Christ-life to birth in the disciples of Jesus, as feminine, in order to complement the masculinity of the Christ's human incarnation. If the Fatherhood of God does indeed symbolise the element of power and moral authority, and if the feminine symbolises tenderness and compassion, then the Spirit whom Jesus, in the Gospel of John, calls 'the Comforter' or 'Consoler' (*Parakletos* in John 14: 16) is aptly thought of as a feminine aspect of the divine.

The whole question is complicated by the fact that some feminists would reject the traditional male–female symbolism of power and authority complementing wisdom and love. If one takes such a view, it would not matter whether the creative principle was referred to as male or female, and calling the Spirit 'she' might be seen as a trivial verbal diversion. It is true that male/female pronouns are meant to symbolise, not genetic physical properties, but different qualities and acts of God. On the whole, however, my own view is that sexual symbolism is deeply rooted in the human psyche, and so it can be more helpful to include a feminine aspect in God than, even grammatically, to exclude it. So I propose to refer to the creative divine Spirit as 'she', even though that is not common in Christian thought.

The Spirit of God is known throughout the Bible as the one who inspires human lives, empowering them with wisdom, skill and strength. The Spirit gives wisdom to the prophets, skill to artists and poets, and strength to warriors and liberators. She touches the human mind with the wisdom of the divine mind, and unites the human will to the creativity of the divine will. The Spirit urges creation towards a fuller embodiment of the divine mind and will, and she involves the divine mind intimately in the emergence of the values of spiritual life in the universe of material energy. It is thus a twofold movement, from the world to God and from God to the world, in which the divine perfections are mirrored in manifold finite forms, by the raising of finite creativity to share in the divine life.

The Spirit of God is not something given by Jesus for the first time, though he does, Christians believe, give the Spirit to the disciples in a new way. But the Spirit itself is universally present, as the dynamic inwardness of God to all finite things. She works in many different ways, though, according to the capacities of creatures and the nature of their relationship to the creator of all, whether known or unknown to them. She is the primal energy of creation, the pressure of God towards goodness. She will never contradict the natures of creatures, and will always be sensitive to their own values and intentions. She is a constant co-worker for good and a constant restraint on evil in all created things, even though her work is often hidden in the complex web of natural and personal causality. Every effective desire for truth, beauty and goodness is prompted and sustained by the Spirit. Every achievement in knowledge, art and virtue is co-operatively caused by the Spirit, and remains forever in God's unforgetful being.

The Spirit searches the depths of God (1 Corinthians 2: 10), finding there the pattern for abundant life. The Spirit brings that life to birth in finite beings, strengthening them, pouring the divine love into them (Romans 5: 5), filling them with hope and joy. The Spirit unites finite lives to God, helping them to raise their minds to God (Romans 8: 26), and conforming them to the divine nature and will. In this work of renewal, co-operation and the transfiguration of life into God, the Spirit follows the pattern of the Supreme Lord, in whom the perfections of this universe are present as in an archetype. The final goal of the universe is that the archetype should become embodied in all the things of the universe, each expressing in its own way a proper part of the infinite fullness of the divine Word.

The Christian vision of God is one of breathtaking scope. It extends to the whole created universe, with its millions of galaxies and its immense solitudes of space. It prompts one to see the universe as a growing, organic whole, moving from a primal chaos, the 'great deep', unconscious, without freedom or feeling, towards a completed and conscious unity with the

Pure Spirit who is the creator of all. The history of the cosmos is, in the Christian perspective, the history of a material universe that is being transfigured, over millions of years, into a perfected sacrament of Infinite Spirit. In this vast cosmic history, the planet earth is a place (perhaps only one of many, or perhaps the only one, we do not know) where the material rises to consciousness of its goal, and becomes able to direct itself towards it – or to turn away from it. The vocation of humanity is to co-operate with God in working towards the full realisation of matter as an embodiment and personal manifestation of the life of God. That full realisation will be in a society of co-operating persons, a 'kingdom of God', but the basis of the Christian hope that it will one day be realised is that the life of Jesus of Nazareth has already embodied the life of God in historical form. At that point in human history, the eternal Word took a human personal life to himself, so that all human lives might one day be united wholly to the eternal Word.

The Cosmic Goal

This account may look unduly optimistic. For the God hypothesis to be fully satisfactory, it has to be true that the third stage, the emergence of a community that fully expresses the divine life, will be realised. From a scientific point of view, we do not know whether or not it will be. Scientific optimists look for a continued increase in knowledge and sensitivity, which will result in a just and peaceful society, free of disease and even of death. Scientific pessimists, however, point out that humans are liable to exterminate themselves at any moment, and that the universe will run down anyway in the end, eliminating any possibility of a final goal of evolution.

Scientific optimists hold that we will be able to control our genetic constitution, thus eliminating the elements of randomness in evolution, with their consequences of disease and death. We will, they say, be able to replace the struggle for survival by co-operation, when we have the technology to ensure abundance of goods for all. Scientific pessimists point out that the people trying to exercise genetic control will themselves be in the grip of the passions of greed and hatred, and are likely to create genetic monstrosities that will probably get wholly out of control. And technology does not make for abundance. On the contrary, it uses up energy resources at an alarming rate, so human conflict over scarce resources is liable to become universally destructive.

Atheists do not have much reason for optimism. It is highly unlikely that blind and indiscriminate physical processes will issue in the existence of a stable moral community of justice and peace. On the theistic hypothesis, however, the cosmic process is not blind or wholly indiscriminate. It involves elements of randomness, but is set up to move inevitably to greater complexity and integrated order. One theistic explanation for the existence of such randomness is that it provides a necessary physical basis for the later development of responsible freedom, which requires that the universe is not completely determined, and so, at pre-rational stages, must be partly random.

It looks as though theists will be optimists, expecting evolution to produce responsibly free and co-operative agents who can subordinate their non-rational and competitive dispositions to more reflective and altruistic intentions. But the story is slightly more complicated. Evolution has produced such agents, but the price of producing them and leaving them free is that they can fail to rise to reflective altruism, and may choose rather to oppress and destroy others than to encourage and sustain them. Theists have to take freely chosen selfishness into account, and that qualifies any over-easy optimism about the future.

In fact, since the earth will be swallowed up by the sun in about five billion years, and the whole universe will eventually run down, it is fairly clear that any enduring goal for the cosmos must lie beyond this physical space–time. I have tried to show that even if all good things come to an end, there can still be a purpose in the cosmic process. But it does not seem possible that a God who cares for creation could let millions of conscious beings die without having had any knowledge of that purpose. If the purpose is frustrated by human evil, it looks as though it might not be achieved, even for a short time, on this planet. If the divine purpose for the cosmos is never fully achieved, or if it is achieved only by a small minority of the conscious beings in the universe, it seems that the divine intention will be frustrated for ever.

These considerations lead to the conclusion that the third stage will probably be beyond the confines of this cosmos, though it must also be the consummation and recapitulation of this cosmic process. Thus, Christians are committed to being trans-cosmic optimists. They are committed to believing that there will exist a community of wisdom, joy and love from which all evil and egoism is excluded. From that viewpoint, the history of this cosmos can be seen as the preparation of sentient beings for existence in the perfected community, though that history has its own distinctive value as the arena of the creative activity of God and the place where created souls can either prepare themselves to share in the life of God or fall into the destructive patterns of egoistic existence.

The Christian will say that the cosmic evolutionary process does not look blind to moral values – it does issue in the existence of moral agents, who can shape the world to good or evil by their own choices. But it necessitates much conflict and suffering, and thus seems, considered on its own, to suggest a God who wills to create goodness through suffering, and is thus far from being a wholly kind and gentle God. We might recall, however, that the biblical God is one who inspires terror, who destroys as well as heals, and whose goodness is more awe-inspiring than tender-hearted. It may be that some modern Christian ideas of God fail to preserve the sense of terror and awe of which the Bible is well aware, and which depict the creator as one who wounds and heals, kills and makes alive, but who in the end offers an overwhelming good to those who persevere.

However, the Christian will also point out that it is a central part of the Christian view that human beings have corrupted the intended process of creation by their free choice of selfishness and grasping desire. As the early theologian Irenaeus said, primitive humans were placed between good and evil, so that they might grow to maturity, whether for good or ill, through the responsible choices they made. In the course of evolution, there must have been a first moment of conscious moral choice. That is the point at which the 'fall of humanity' began and humans were estranged from that natural fellowship with God which should have been theirs, and from their natural ability to relate unselfishly to one another.

The New Testament writers see God's purpose for creation as having been corrupted by the doomed attempt of sentient creatures to turn away from obedience to God. The created cosmos falls away from its intended goal, and becomes subject to futility and spiritual death. The wages of sin is death, existence in estrangement from God, the only source of life. Within this estranged world, the divine Word takes human nature in order to heal the sickness of self, manifest the divine love, and lead humanity back to its proper destiny. When the first Christians looked for the *parousia* of Christ, the manifest presence of the eternal Word, they were looking for a future in which all creation would be renewed in the divine Word, restored to its predestined goal. In that future, Christ will be manifest as the pleroma, in whom all finite realities will be fulfilled and brought to their proper perfection. The cosmic goal will be achieved, since God has the power to ensure it, but it will transcend the history of this cosmos. The process of cosmic history and the struggles, failures and endeavours of creatures within it will be what gives significance to the realised goal. The cosmic process as a whole is what realises the purpose of God, but only the final appearing of Christ in glory, and the final uniting of all things in Christ, invests the process with its true meaning and significance.

3

Sacred Cosmology: the Genesis Creation Narratives

The First Creation Narrative

Creation is not, properly speaking, just the origin of this universe. It is the universe as a whole, from beginning to end, in its dependence upon and interaction with the creator God. Nevertheless, Christians do believe that this universe, and all the universes there may ever be, have their origin in God. Modern cosmology leads one to accept that this universe has an origin, a first moment of time. But in other respects it gives quite a different account of the origin of the universe from the Bible. What the Bible provides is a sacred cosmology, a spiritual interpretation of the universe's origin, nature and destiny, not a scientific cosmology. Some people think that the first two chapters of the Book of Genesis are to be interpreted as scientific accounts of the beginning of the universe. If this were the case, there would be a definite conflict between the scientific worldview and the biblical worldview. However, the idea that the Genesis origin story is a scientifically accurate account, or that it is meant to be a primitive attempt at science, which happens to be incorrect, is completely misconceived.

Most tribal religions have stories of the origin of the world. For example, the Cheyenne native Americans have a story of the creation of the world by the primal spirit Maheo, who, having first created a great salt lake, with the help of a little coot shaped the land out of a ball of mud on the back of Grandmother Turtle, and formed humans out of his own ribs.

I spoke to a tribal elder, who said that it would never occur to Cheyenne to think of this as a literal account. It is a story that expresses important spiritual truths – the necessity of co-operation between creator and humans, the need for proper reverence and care for the earth mother and for all living creatures, and the importance of the interdependent unity of male and female, for example. Spiritual truths are depicted – truths about how humans ought to live in relation to God, to their world and to others – in the form of a story. To ask whether this story ever happened in history is quite simply to miss the whole point.

So it is with the two distinct creation stories in Genesis. They come from different tribal origin stories, and they have been placed side by side by the editor of Genesis, because they express different, important, spiritual truths. The editor was not worried by the fact that, taken literally, they are incompatible – in the first story (Genesis 1: 1–2: 3), God creates vegetation, then fish and birds, then animals and finally human beings, whereas in the second story (Genesis 2: 4–25), God creates a male human first (before there was any vegetation), then trees, then animals and birds together, and finally a female human. This was not worrying, because different stories can convey different spiritual truths. Like metaphors, they cannot contradict one another – only literal accounts can be contradictory – but they may very well complement one another. The two Genesis origin stories offer important complementary spiritual truths, and the editor did not wish to lose any of those truths. So he set the two stories side by side, making it quite clear that they were not to be taken as literal history, but as different narratives, each of deep spiritual significance.

The first story begins with a formless dark void, *tohu wabohu* – the waters of the 'great deep'. The sea is the symbol of chaos, the formlessness that always threatens ordered creation. But over that chaos sweeps the spirit of God, the divine breath or wind (*Ruach*) which has creative power, and can bring form and order into being. The first act of God is the pronouncing of a word, 'Let there be light.' Words are the expressions of thought and intention. So the world comes into being as the expression of a divine purpose. But God does not say, 'I create light.' God says, 'Let there be light.' God, the story seems to say, does not wholly determine or invent what shall exist. Rather, God permits light, which exists as a possibility, to exist. We might think of light, and of all creation, as existing potentially in the divine, waiting for permission to exist in actuality. God's role is to draw back the veil that prevents things existing, to permit them to be, and to do so by pronouncing a word of liberation. So things are released into the expression of their own power. The Word does not impose a nature on things from outside, as it were. The Word of God releases things to express their own inherent potency.

The first potency to be released into actuality is light, a light that exists before sun, stars or moon. That light is the glory of God itself, appearing in primal chaos, released into being by the Word and energised by the Spirit. In 2 Corinthians 4: 6 this light is, from a Christian viewpoint, explicitly associated with the Word, the eternal Christ: 'the God who said, "Let light shine out of darkness" has shone in our hearts to give the light of the knowledge of the glory of God in the face of Jesus Christ'. John's Gospel affirms that Christ is 'the true light, which enlightens everyone' (John 1: 9). And the Book of Revelation asserts that the new Jerusalem will need no sun or moon, since 'the glory of God is its light' (Revelation 21: 23).

It would totally miss the point to ask whether light really was the first thing God created, in a literal chronological sequence. It would miss the point just as much to claim that modern cosmology confirms that light (in the form of photons) was the first created reality. When Genesis speaks of 'the first day', it speaks of the first of the divine creative acts, in order of logical priority. The light of divine glory, the primordial fire, manifests in the element of water, the primal chaos. The divine Word permits it to manifest, and the divine Spirit divides the light from the darkness, the order of glory from the anarchy of darkness. 'God saw that the light was good' – the manifested glory of God is a new reality of great value, in which any consciousness, including that of God, can and should delight.

The second great creative act of God is the conquest and ordering of chaos. The 'separation of the waters' of the great deep is also referred to in the Hebrew Bible as the defeat of the chaos-monster: 'You divided the sea by your might; you broke the heads of the dragons in the waters. You crushed the heads of Leviathan' (Psalm 74: 13 and 14). The prophet Isaiah writes: 'Was it not you who cut Rahab in pieces, who pierced the dragon?' (Isaiah 51: 10). The dragon of chaos is subdued, but will not finally be killed until this creation ends, and a new creation comes into being: 'On that day the Lord with his cruel and great and strong sword will punish Leviathan the fleeing serpent, Leviathan the twisting serpent, and he will kill the dragon that is in the sea' (Isaiah 27: 1). The dragon-slaying and the dividing of the deep symbolise the divine ordering of primal chaos, which stands as a continuing, but restrained, threat at the borders of creation. A space is cleared, the realm of air, the third element, the realm where the Spirit blows free and begins her creative work.

The third divine act of creation is the emergence of the realm of earth, the last of the four elements. The earth becomes the mother of life, and God gives her freedom to bear from herself (a maternal verb) all plants and trees, to become a living entity, a biosphere within which conscious life can take shape.

The fourth divine act is the ordering of time. The 'lights' of the sun, moon and stars are not divine beings or even angels, but 'signs' of the glory of God, giving rhythm to seasons and years. The sun gives light to the earth, but though many have worshipped the sun as a god, for the Hebrew story the sun is simply an instrument of the true God. It is destined to pass away, when the purpose of creation is fulfilled: 'They need no light of lamp or sun, for the Lord God will be their light' (Revelation 22: 5). Human life is lived out in measured time, and the sun and the stars lay down the temporal pattern that is necessary for the implementing of human goals and intentions.

The fifth divine act is the emergence from the ordered waters and the filling of the realm of air with living creatures. Water and air are filled with life, and that life is blessed by God, so that it flourishes and multiplies. Even out of chaos, God draws forth life which flourishes, which is creative and which thus to a small degree shares in the creative activity of God. That is the meaning of 'blessing' (Genesis 1: 22), a raising by God of elemental matter to a share in the divine creativity, a sharing in which true welfare and fulfilment is to be found.

The sixth divine act is the bringing forth from earth of the mammals, in which for the first time the power of reflective awareness and deliberative movement is found. Among these animals, God created humankind, in the 'image and likeness' of God. As an image in a mirror reflects the reality, so humans are meant to reflect the nature and activity of God on the created earth. God creates, orders and blesses (fulfils). So humans are to create – they are to be fruitful, not only biologically, in having children, but more importantly in bearing the 'fruit of the Spirit': love, joy, peace, patience, kindness, generosity, faithfulness, gentleness and self-control (Galatians 5: 22). In their creativity they are, like God, to 'let things be' in their true natures, not to use, or misuse, things for egoistic ends of human pleasure.

Humans are given 'dominion' over all living things (Genesis 1: 28). They are to exercise the just and gentle rule of love, which allows living things to flourish. At this point animals are forbidden as food (1: 29), and the dominion humans are to exercise is a stewardship of care, not an unlimited right to do as they please with God's creation. So, while they are commanded to 'fill the earth and subdue it', they are also reminded by God that 'the land is mine; with me you are but aliens and tenants' (Leviticus 25: 23). Since God has created all things, they are never 'owned' by humans, whose task it is to help make the world fruitful, to bring all things to their proper fulfilment. Humans are to do this in partnership with God, since they are given the gift of being 'like God', of being able consciously to know and love God as father and even as loving partner.

With the emergence from primal chaos of beings capable of conscious relationship with God, the work of creation is finished. At that point, at least on this planet, history begins, the story of the human pursuit of the divinely intended goal, its loss and recovery, and its final achievement. God continues to interact with humans, in calling, judgment, forgiveness and promise. But the great work of calling the elemental structures of the universe into being is complete. The first origin story therefore concludes with the picture of God 'resting' on the seventh day, the *shabat*, and so setting every seventh day apart as a holy day, a day free from toil, a day free to contemplate the works of God, the beauty of creation, and delight in it. *Shabat* is taking time to delight in God and all the works of wisdom, to share in the contemplation of all that has been created, with God who, having created all things, 'rested and was refreshed' (Exodus 31: 17).

The Second Creation Narrative

The second creation story uses different narrative symbolism to convey four main spiritual truths. First, humans are formed 'from the dust of the ground' (Genesis 2: 7), enlivened with the 'breath of life'. They are not pure immortal souls, fallen into the material world, as in some religious views. They are bits of matter, and their life depends entirely on God, who gives it and takes it away ('When you take away their breath, they die and return to their dust': Psalm 104: 29). The Bible does not look for a release of the soul from the material world. It looks for the creation of a body which is fitted for mediating the Spirit of God within the material world.

Second, the true home of humanity is a garden of delight (Eden), in which all things can flourish in total dependence on

God. At its centre is the tree of eternal life, watered, as the New Testament adds, by the 'river of the water of life, flowing from the throne of God and of the Lamb' (Revelation 22: 2). For Christians, the words of Jesus come to mind: 'The water I give will become . . . a spring of water gushing up to eternal life' (John 4: 14). The intended destiny of humanity is, by continuing dependence on God, to live forever 'in the fullness of joy' (Psalm 16: 11).

Third, humans are 'to till and keep' the garden. They have the responsibility of caring for it, and eventually of building within it the city of peace (Revelation 21: 2), a community in which new values can be created, shared and appreciated. That city will be built by wisdom, yet it must be the wisdom of *kenotic* (self-giving) love, manifested by 'Christ crucified' (1 Corinthians 1: 23 and 24). Christ, 'though he was in the form of God, did not regard equality with God as something to be exploited, but emptied himself . . . and became obedient to the point of death' (Philippians 2: 6–8). Wisdom directed by love is self-renouncing, and it allows God alone to rule. On the other hand, wisdom without love, used for the sake of gaining power or wealth, leads to self-destruction. It is such wisdom directed by power that is to be destroyed by the love of God: 'I will destroy the wisdom of the wise, and the cleverness of the clever I will thwart' (1 Corinthians 1: 19). Wisdom is symbolised by the 'tree of the knowledge of good and evil', which stood in the garden and which God prohibited until Adam had fully matured in the love of God. When the serpent, another representation of the dragon of chaos, urged humans to 'eat the forbidden fruit', to claim autonomy, knowledge and power, humanity brought destruction upon itself by claiming knowledge without love and without responsibility, in the name of selfish desire. The result would be 'ejection from the garden', loss of the sense of the presence of God, and a slow but inevitable spiritual death.

Fourth, God made Adam 'a helper and a partner'. To be human is to be in relationship to others, and true personhood can only be found in community, not in the Neoplatonic 'flight of the alone to the alone'. Adam named the animals – that is, he had the responsibility to understand and control them. But a full moral and mutually responsible relationship only exists on earth between members of the human species. With the creation of Eve, a special relationship is established between men and women so that they can live together in the deepest form of union, an unbreakable relationship of loyalty and trust, a union which mirrors that of God and Israel ('"On that day", says the Lord, "you will call me 'my husband'"': Hosea 2: 16).

The first creation story is primarily concerned with the ordering of the elements of the universe, as beings with moral and rational awareness emerge from primal chaos. The second creation story focuses more specifically on human nature, as properly material and relational, and as finding its fulfilment in total dependence on God and in the self-giving wisdom of love. Together, these stories give great insight into the nature of humanity in a universe created by a wise and powerful God. They do not provide a scientific account, but are inspired stories conveying spiritual truths in narrative form. Modern scientific accounts of an evolutionary universe complement these narratives by sketching out the wider cosmic perspective into which their underlying teachings can be fitted. The fit is, I think, a natural one, so that the scientific view and the biblical view together provide an intellectually satisfying and spiritually illuminating account of creation.

4

Explaining the Universe

Creation and Scientific Method

The difference between the traditional religious pictures of the universe and the modern scientific picture is not so great, after all. Many scientists may not see human beings as intentionally created by God. But many others do, and the scientific picture itself is consistent with the idea that humans have been created because of their ability to realise intrinsic values, and especially the value of knowing and loving God. Both scientists and theologians – often, of course, they are the same people – can see the universe as exhibiting amazing beauty and wisdom, as being the sort of universe that a being of great power and wisdom could create. And both can see the universe as having a cosmic goal, which lies in the consciously apprehended fulfilment of the potentialities implicit in its origin. The Christian faith sees Christ as the archetype, the model, for all the forms of beauty and wisdom in the universe, and as the pleroma, the goal and fullness of being, towards which the cosmic process moves. In human history, Christ has taken embodiment in human form in Jesus of Nazareth, so that we can have access to the splendour of the cosmic Christ through the humanity of Jesus. But it is when we see Christ as the eternal Word, archetype and pleroma of creation that we begin to discern the grandeur of the New Testament vision. Far from being undermined by science, this vision is even further deepened and expanded by contemporary cosmology.

In fact, the relation of modern science to belief in God is even closer than these points might suggest. There is one fundamental dogma in modern science that cannot be proved, but without which science cannot exist. It is the dogma that there is an explanation for everything. If anything exists for which there is no explanation, science is stumped. If rabbits, flies and top hats just started appearing in the universe for no reason at all, just coming into existence out of nowhere, physical science would virtually break down. One would never know just when a rabbit was going to appear in the laboratory and ruin all one's carefully planned experiments. The irrational, the inexplicable, the totally random, puts an end to science.

This belief that nothing happens without some explanation, some reason or cause for its existence, has its roots in religion. Or rather, one of the roots of religious thought is just the same as this root of scientific thought. Science and religion are united at this very basic level. Isaac Newton remarked that his search for simple underlying laws of nature was prompted by the belief that a wise creator would have designed the universe to run on such simple principles. The reason events happen in intelligible, largely predictable ways is that they act in accordance with general principles, laws of nature. The laws of nature look just as if they have been selected as the most simple and elegant principles of intelligible change by a wise creator. Belief in the intelligibility of nature strongly suggests the existence of a cosmic mind that can construct nature in accordance with rational laws.

Appeal to the general intelligibility of nature, its structuring in accordance with mathematical principles that can be understood by the human mind, suggests the existence of a creative mind, a mind of vast wisdom and power. Science is not likely to get started if one thinks that the universe is just a chaos of arbitrary events, or if one thinks there are many competing gods, or perhaps a God who is not concerned with elegance or rational structure. If one believes those things, one will not expect to find general rational laws, and so one will probably not look for them. It is perhaps no accident that modern science really began with the clear realisation that the Christian God was a rational creator, not an arbitrary personal agent, who leaves many events to the whims of angels and demons and interferes in unpredictable ways every now and then. The best breeding ground for science, in its modern sense, is the idea of one God who creates the universe on principles of wisdom and reason. This is exactly the Christian idea of God, a God who does not create through arbitrary or irrational acts of will, but who creates through the Logos, the Word, the principle of true reason or wisdom ('Christ is the Wisdom of God': 1 Corinthians 1: 24). It was, however, not until the High Middle Ages that Christian theologians were knowledgeable and self-confident enough to begin to see the implications of this idea of God.

If one believes in such a creator, one will be able to proceed on the assumption that the mind may discover the basic structures of nature if it works on principles of true reason, seeking some rational explanation for the occurrence of every event. This is just what modern science does, and its studies have proved remarkably successful. The assumption of one rational and wise creator is in this way strongly confirmed by the success of science.

Creation out of Nothing

Explanations in science usually work by showing how complex and often seemingly chaotic processes are the result of the operation of general laws on simpler elements. Thus, the transformation of water into steam is explained by the increased motion of the molecules that make up water, due to the application of heat. We explain a change from one state to another by referring to smaller constituents of matter, atoms and electrons, and the general laws of their interaction. In recent decades, we have seen how the structure of atoms can, in turn, be explained by reference to even smaller particles, quarks, and how general laws of interaction can be subsumed under four basic natural forces. Some physicists search for an ideal 'Theory of Everything', which would unify these basic forces under one general law of interaction that might even explain how quarks arose in the primal energy-exchanges of the early universe. In quantum cosmology, quantum theory and relativity theory are both used to try to show how this universe might be generated by quantum fluctuations 'in a vacuum', where all electrons are in their ground state.

Some physicists speak of this as 'creation out of nothing', but this is a misleading expression in two main ways. First, the word 'creation' is misused by such writers, because they are talking about the *origin* of the universe, the first microseconds of its existence, not about the creation of the universe by the intentional action of God. It is true that Christians have usually used the term 'creation' to refer to the beginning of this space–time universe. But they have only used the term to refer to a certain sort of origin, an origin which is due to the intentional act of a divine being. To 'create' is to make something exist, knowing what it is going to be, and intending to make it. The simple fact that the universe may have started with the Big Bang does not mean that it was created at all. It may have just started, or come into being in some other way. For Christians, it explains the origin of the universe well to say that it was created, but that has to be established.

When Christians, and most other believers in God too, talk about creation, they do not just mean that God started the universe going. They mean that the whole of space–time, from beginning to end, depends in every detail on the conscious and purposive act of God. One can believe the universe is created even if it never had a beginning, a first moment of time, and even if it will never have an end. As a matter of fact, most physicists think that our universe has both a beginning and an end. But that fact in itself does not make creation more or less likely. Augustine thought, long ago, that God might create other universes before and after this one, so time might never have begun. But all time would still be created, purposively made, by God.

It is also wrong to think that God might have started the universe going, and then left it to fend for itself. The universe could not exist even for a second without God, because on the hypothesis of creation the whole universe, from beginning to end, and everything in it, depends for existence on God. There is no creation if there is not a God who intends created things to exist and, for most believers in God, only God is uncreated. Everything else is brought into being by a conscious, intentional act of God.

The universe might, according to some cosmologists, just come into existence by chance, for no reason, or it might do so by some sort of inner necessity, because it has to do so. That would be an origin of the universe, but it would not be creation. Those cosmologists who say that the universe just comes into being out of nothing, without being intended by any God, vacillate between chance and necessity as the origin of the universe. They often envisage a sort of fluctuation of energy, in accordance with basic quantum theory, which produces successively all sorts of configurations of space, time and fundamental forces. This fluctuating process just happens on its own, without direction or purpose, necessarily running through all the possible combinations allowed by the theory. Out of these fluctuations, sooner or later it is bound to happen, by chance, that a coherent universe like this one is generated. So this universe is generated by chance, not intentionally. But it will necessarily be generated sometime, as the quantum fluctuations run through all the possibilities open to them.

In this way, it might be said that the universe does not need a creator. It emerges by a combination of chance and necessity, 'out of nothing', blindly and without purpose. This, however, is the second misleading way of speaking about the origin of the universe. 'Nothing' is lack of everything, complete emptiness. But on this sort of quantum cosmological theory, this space–time emerges, not from complete emptiness, but from a very full and complex set of quantum fields, constantly fluctuating in regular and systematic ways – so that every possibility is realised in time. Where do the laws of nature that regulate the fluctuations of quantum fields come from, and in what sense do they really exist, even when there are no material particles for them to order? Where does the primal energy that acts in accordance with quantum laws come from? And what ensures that every possible state will sooner or later be actualised?

There is no intrinsic likelihood that every possible state will sooner or later be realised. In fact, there is absolutely no reason why the whole array of possible states should ever be realised at all. Just one or two states might repeatedly be realised, while all others remain forever unrealised. If all possibilities are to be realised, there needs to be some law which ensures that this will be so. But how could one be sure that such a law would continue to exist, or that events would actually occur in accordance with it? It seems that all the questions are being begged, if all this is put down to pure chance. Proponents of such a view simply have not considered how radical a 'pure chance' explanation really is. It allows absolutely anything to happen, for no reason at all. So, in fact, it precludes all events being guaranteed to follow some law of successive realisation. For on the pure chance theory the laws themselves could change or cease to exist at any moment. Then, of course, no one could be sure that every possibility would be realised in accordance with some general law. So the 'quantum fluctuation' view fails to justify the belief that this universe is bound to be realised sometime. We have to return to the hypothesis that it just does exist by chance, that is, for no reason at all, and without any possible explanation.

Such a hypothesis amounts to a rejection, at the last moment, of the quest for intelligibility that is the very foundation of science. It seems odd if science ends with the final rejection of its own most fundamental dogma. It might be said that these are the ultimate brute facts, the simplest possible set of consistent laws and energy states that could exist. Once we have got down to this simplest possible level, we can go no further. We have reach the bedrock of all possible explanation. When we try to explain a phenomenon, we try to do so in terms of simpler elements and more general laws. When we get down to ultimate simplicity, we can do no more. Explanation, it might be said, has to stop somewhere, and this is it.

Many physicists are dissatisfied with such a disappointing result, and rightly so. If only, they think, there was some absolute explanation, something that explained its own existence as well as the existence of everything else. What would such a thing be like? It would not be just a contingent thing, something that happened to exist, but which could easily have not existed. That would need an explanation of why it exists. It would have to be something that has to exist, something to which there is no alternative. That is what philosophers have called a necessary or self-existent being, a being that exists by its own nature, is not dependent on anything else, and does not simply exist by chance.

Some physicists think that perhaps there is a necessary being that is some amazing mathematical equation. Mathematics, after all, is necessary, so it fulfils that requirement, and it might somehow, inevitably, give rise to the physical universe. Yet it is very hard to see how mathematical equations can exist just on their own, and even harder to see how they can give rise to a physical universe. This is where theism has the advantage. For theists, God is the one and only ultimately necessary being. If there are necessary mathematical equations, existing even before this universe came into being, the obvious place for them to exist is in the mind of God, the supreme cosmic intellect. It is God who can select a world from all the available possibilities, and so God's creative choice can explain how a physical universe comes from a set of ideas, mathematical or otherwise, in the mind of God. For theists, God, the cosmic intelligence, is not just a contingent, accidental, reality. God is the one and only necessary being. God cannot fail to exist and be the general sort of being God is. That is the one huge difference between God and every created thing, and it enables God to be the finally satisfactory explanation of the existence of the universe in a way that nothing else could be.

Many quantum cosmologists thus agree with theists in postulating a necessary being with a certain sort of ultimate simplicity, out of which this universe arises in a non-arbitrary way. The crucial question is whether the universe arises purposively, by knowledge and intention, or blindly, by chance. The common faith is that events have an intelligible structure, that the quest for explanation should be pushed as far as one can go. The ideal explanation, both agree, would be one that left nothing unexplained, not even itself. But if that cannot be achieved, analysis might at least get to a level of irreducible simplicity and generality, where no possible further explanation could be given. That is the final aim both of fundamental science and of fundamental theology. They are not so far apart after all.

The Choice Between Chance and Design

It is a very remarkable and unexpected fact about the universe that all its ordered complexity results from a cumulative construction out of ultimately rather simple principles. What is remarkable is that there should be such a simple structure which gives rise, through a cumulative and ordered organisation, to a level of complexity rich enough to generate such things as consciousness and free action. It is not at all likely that simple structures should generate ordered complexity in this way. The fact that they do so strongly suggests intentional design, rather than blind chance. It could be due to chance, but it is much more likely if it is due to design by a powerful and wise intelligence. This can be shown fairly readily.

There are a virtually infinite number of possible universes, with many possible combinations of kinds of elementary particles, basic laws and values like Planck's constant and the gravitational constant. Most physicists think that virtually none of them would give rise to a coherent and continuously existing universe, and perhaps only this one could give rise to rational sentient life. It is hard to show that there is only one possible universe that could give rise to rational life. Nevertheless, the physical conditions necessary to produce life in this universe will occur in very few of the huge number of possible universes there are, and it could well be true that human life, life very like ours, could only exist in a universe whose basic parameters were as ours are.

If our universe is just one out of a virtually infinite number, then its existence is almost infinitely improbable. Of course, it is no more improbable than the existence of any other possible universe, however short-lived or abortive, so we cannot say it is more improbable than other universes. But it is still hugely improbable that it should exist at all. If, however, we suppose that there is a God who creates a universe in order that conscious life should come to exist, then it is much more likely that this universe will exist than most other possible, abortive, universes. Moreover, given the existence of this universe in its earliest stages, it is immensely more likely that conscious life will evolve in it than not. Indeed, if God wills to create sentient life through a process of ordered evolution, it is virtually certain, it has a probability very near to 1, that this universe, or one very like it, will exist.

Given the hypothesis of God, the existence of a universe like this is really quite likely, or not at all surprising. Given the existence of a universe like this, it is virtually certain that sentient life will come to exist in it. Thus, on the hypothesis of God, the existence of this universe is quite probable, whereas on the hypothesis of chance, its existence is almost infinitely improbable. Since we should always choose the hypothesis that raises the probability of the facts it is posited to explain, it is more reasonable to think that the universe results from intentional creation than that it originates by blind chance.

5

Theistic Explanation

The Necessity of God

The hypothesis of God is a good explanation for the universe, since it makes the existence of the universe much more probable than it would otherwise be. But of course that will only be a convincing argument if the existence of God is not itself very improbable. If God was just as improbable, or even more improbable, than the universe, it would be useless to appeal to God to explain the universe.

So is the existence of God very improbable? The answer must be 'yes', if one thinks of God as a very complex person, who just happens to exist, and who decides to create a universe as a result of some arbitrary whim. Such a God would be as improbable as the universe God is supposed to explain, and since his reasons for creating the universe would be inscrutable, they would not explain why the universe is the way it is at all. It follows that God should not be thought of as a very complex person.

What, however, is the alternative? In fact, the classical Christian idea of God offers a very attractive alternative. It may seem rather abstract and difficult to understand, but it repays careful study, since it is a useful corrective to the popular (mis)understanding of God as an external interfering person. The classical Christian belief is that God exists by necessity, that is, with a probability of 1. What makes any actual being improbable is that there are many alternative possible beings.

The existence of just one being out of millions of alternatives is very improbable, though no such being is more improbable than any other.

What would eliminate the improbability is the existence of a being to which there were no real alternatives at all, a being that could not fail to exist. Such a being is beyond human comprehension – as surely one might expect God to be – but I think we can at least dimly grasp the idea of such a being. It is the idea of a cosmic intelligence that is omniscient, that always knows every possible state of affairs there could ever be. In fact, possible states only exist in so far as they exist in something actual, in so far as they are in the consciousness of God. If God was not thinking of them, they would not exist, even as possibilities. So if God did not exist, there would be no possibilities. Nothing ever would be possible, and if nothing was possible, obviously nothing would ever be actual. It looks as though, since many things are actual, God must exist. Furthermore, since these things must always have been possible, God must exist anyway, even if there is no actual universe. There is no alternative to the existence of God. That is how God can be a necessary being. God is the actual being in which all possibilities exist, the ground of all possibility, which cannot be thought away without destroying every possibility of being.

Perhaps I should just point out that this does not prove that God exists, as if by some conjuring trick of logic. It just says that we can get a vague idea of a necessary being by thinking of a being in which all possibilities exist, precisely as possibilities. If the idea of such a being is coherent, then we can see how an omniscient cosmic intelligence could be the necessary being that provides an absolute explanation for the universe. What we cannot prove is that the idea is coherent – some people would think that the idea of possibilities somehow existing is too odd. Nevertheless, from Plato onwards many philosophers have considered that 'possible worlds' exist in some sense. To do what St Augustine did, and put them into the mind of God, seems plausible, and it seems to be a coherent idea. The point is that God is not just one more being who happens to exist, and who needs explaining just as much as the universe does. If God really is a necessary being, then God is self-explanatory, and the explanation is that, if we really understood what God was, we would see that God could not fail to exist. God has to exist, if anything at all is possible.

We can now think of God as a being who conceives the idea of all possible states, as an omniscient mind. This mind is complex, in containing an uncountable number of ideas. But it is simple, in being one mind, containing an exhaustive and necessary set of ideas. God is being postulated as the cause of this universe. God selects from the total set of ideas a particular subset, and gives them actual existence. Why should God do this? There is one fully intelligible reason for choosing something to exist, and that is to choose something for the sake of its goodness, or value. Some possible states are of more value than others, and it is rational to choose possible states of more value and cause them to exist. The best explanation for the existence of the universe is that it is selected for the sake of its goodness from the total set of possible universes that exist in the mind of God.

Thus, the simplest explanation of a complex contingent universe is that it is given existence by just one being that exists by necessity and causes the universe to exist by choice, from a necessary set of possibles. God is not a very improbable being who just happens, by chance, to exist. God is a being that, uniquely, exists by necessity. If one is able to have some grasp of this idea, one begins to see the sublimity of the Christian idea of creation, and the way in which the cosmos depends on the existence of a self-existent God.

Creating a Contingent Universe

God's existence is explained by necessity. The universe is explained as chosen for the sake of its goodness by God. Some philosophers have thought that there is one possible world which is the best of all possible worlds, so a rational God is bound to choose it. That is Leibniz's hypothesis, and it is a good one. Except that there is not just one best possible world. There are many very different sorts of good things, which cannot be measured against one another on a common scale. Is a universe containing more Beethoven symphonies better or worse than a universe containing more Mozart sonatas, or more snowy mountains, or more football games? The question is unanswerable, because one simply cannot compare these good things. So if there are many good possible worlds, which cannot be compared with one another on a common scale, there is bound to be an indeterminacy of choice.

What Leibniz failed to see is the true contingency of the universe. This universe does not have to exist. It is only one of many possible universes. Though it contains sorts of goodness that other universes do not contain, it also lacks many goods that other universes would contain. It is not the best possible universe, and unlike God, it does not have to exist. If one believes that there is a creator of a truly contingent universe, then the act of creating that universe must also be contingent, it must be a matter of free choice. If God creates this universe, God does something that God does not have to do. God could have acted otherwise, could have created a different universe, or not created any universe at all. That means that though God exists necessarily, God must have the power to act freely. This idea of God as having some necessary powers – existing, having the power to create, and knowing all possibilities, for example – and as having some contingent powers – the ability to create a contingent world, for example – is a perfectly coherent idea, but it did not occur to Leibniz.

The nearest parallel to this idea is in quantum physics. The fundamental particles of quantum physics have some necessary properties. They cannot just all disappear at once or turn into quite different sorts of particles at random. They necessarily obey the general laws of physics, such as the conservation of energy laws. Yet their behaviour is not necessary in all respects.

According to most interpretations of quantum theory, there is an area of indeterminacy, within which the behaviour of particles is not completely determined by any laws or initial conditions. So fundamental particles are partly governed by necessity and partly contingent in their behaviour. This is only an analogy for a God who is necessary in existence and general character, although contingent in many particulars, but it may help to show that the idea of a God who is both necessary (in some respects) and free (in others) is an entirely coherent idea.

It is just the idea that is required for a free creator God. So we can say that God exists by necessity, necessarily conceives all possible worlds, and contingently selects one for the sake of its distinctive goodness. God is a being who unites the apparently contradictory powers of necessity and freedom. This is another very important sort of simplicity. The divine necessity defines unchangeably the nature of God and the possible worlds which God can create. The divine freedom selects an actual world by an act of sheer creativity. The insight Leibniz did not have is that creativity is a great intrinsic value, something worth actualising just for its own sake. Indeed, one might expect that if God is the creator of all, it would be obvious that creativity is a great good. As human beings are created in the image of God, and called to shape their lives in the likeness of God, one of the greatest values in human existence also will be the value of free creativity. That will have major implications for the sort of universe a God who creates beings in the divine image will create.

If the universe is such as to generate freely creative beings, its principles of necessity (the laws of nature) must be non-deterministic. There must be room for creative freedom to operate. That means that the laws of nature must not determine every outcome in a predictable and unchangeable way. And that means that, even before rational freedom comes to exist, many natural processes will have the appearance of randomness or indeterminacy. This will not be total randomness, though, since it will be finely tuned to produce just that degree of indeterminism that will allow free actions to come to exist within a suitably developed physical structure.

Does that mean that the exercise of such free and creative powers cannot be explained, either in humans or in God? They cannot be explained in terms of determining or necessary causes, by definition. But they can be explained by final or axiological causes, that is, in terms of their goals or purposes. God acts for the sake of goodness. One intrinsic good is creativity, the sheer creation of the new. Creative action is a good, and it excludes determining causality. God is therefore the freely creative cause of the universe, not a being who has to create exactly this universe out of necessity.

The quest for explanation is, in principle, at an end. In principle, but not in fact, for humans will never, on earth, have the intimate understanding of God that would actually enable them to explain why everything is as it is. But they can understand enough to know that there is such an explanation. God knows what it is, and we know that God knows. That makes a tremendous difference, for it means that nothing in the universe is arbitrary, that the whole cosmos is intelligible, a work of divine Wisdom. But it also prevents us from claiming that we can actually provide the final explanation for everything. Only the Holy Spirit can search the mysteries of the divine mind.

We can say, though, that the universe is best explained as generated by a wise and powerful God for the sake of the distinctive sorts of goodness it alone can realise. God is explained as a being who exists by necessity, as necessarily all-knowing and capable of creative action, and who acts freely and creatively to bring about some universe of finite creatures because such action is itself a great good. On such a hypothesis, the probability of the existence of God is as high as it could possibly be. That is, since it is not possible for God not to exist, the divine existence is certain. The probability that a universe producing sentient beings will exist, if any universe at all exists, is extremely high. So the hypothesis of theism must be given a very high probability, as long as it is coherent.

Is God a Hypothesis?

It may be objected that to talk of God in this way, as if God was a hypothesis with a certain probability, is just to mix up science and religion in an unhelpful way. I must admit that I was quite surprised myself to find that God was emerging as a highly probable hypothesis for explaining the universe. I was brought up, philosophically, on a diet of post-Kantian philosophy, and it was almost a self-evident truth that the existence of God could not be established by reason. To believe in God was to commit oneself totally, not tentatively to accept some hypothesis.

Well, as Professor Joad used to say, it all depends on what you mean by 'hypothesis'. In the sciences, one would expect a good hypothesis to predict some results that could be experimentally confirmed. One might devote a lot of time to trying to disconfirm a hypothesis, or construct more elegant alternatives that might explain a wider range of data. The trouble with the God hypothesis is that it seems to explain almost anything. Nothing can disconfirm it. And believers seem almost irrationally committed to it – or at least they believe it with a strength far beyond what the evidence might suggest.

God is being proposed as a hypothesis that 'explains' the universe, in that, if we could understand God, we would see just why the universe exists as it does. The universe would not follow from God, as if by some general law, because of the personal creative element in the creation of any universe. In this sense, theistic explanation is more like historical explanation, in terms of purposes and desires, than like the general-law explanation common in the sciences. But one can still explain the existence of the universe by attributing it to an intelligible choice, even though such an explanation is unlikely to get into physics textbooks.

And there are some consequences of the hypothesis. Any universe created by God must be overall good, and all the evil in it must be a necessary condition or consequence of some process that produces a great good. Any such universe will be structured on elegant intelligible principles, and contain communities of responsible conscious beings. So the God hypothesis could be disconfirmed if, for instance, there was irredeemable evil, anarchy or lack of purpose and pattern in the lives of beings in the universe. Producing a decisive disconfirmation is virtually impossible, since the theist always has life after death to fall back on. But there are certainly relevant observations and experiences that will tend to confirm or disconfirm the hypothesis.

Many scientists may well be uninterested in a hypothesis that is purely hypothetical (*if* we could understand God, we *could* explain . . .), largely historical (in terms of unique and unrepeatable choices), and untestable by specific experiments. In that sense, God is unlikely ever explicitly to figure in scientific theories. Yet others, particularly cosmologists, do feel the attraction of an 'absolute explanation', even if they can never get the details of it. One might call this a 'meta-scientific' explanation, since it seeks to explain why the whole structure of the universe is as it is. At that level of abstraction, I think God is the best explanation on offer.

The religious believer may still rightly feel that this is a very abstract and tentative idea of God. After all, the universe may simply have no absolute explanation. The believer's faith is based on personal experience and commitment, not on tentative speculation. Nevertheless, the believer has an interest in knowing, not only that the object of belief, God, is consistent with the best scientific knowledge, but that in fact it can make a good claim to be the most reasonable belief there is, in the light of modern scientific knowledge. There will continue to be disputes about God, precisely because there is no neutral ground from which 'pure reason' can come to definitive conclusions about such ultimate matters. Yet faith in God is not a blind leap in the dark. The hypothesis of God is one that would make the existence of such a universe as this much more probable than it would otherwise be.

It is also the simplest possible hypothesis that is comprehensive enough to account for all the complexity of this universe, including the facts of consciousness, value, purpose and personal existence, as well as the facts of physical order and complexity. It posits just one being, with one ultimate reason for creating (for the sake of distinctive value), capable of being defined by one simple formula. For God, who knows all possible states, and who has the power to actualise these states, will ensure that the divine being itself contains the greatest number of actual values at any time, and so God will be, in Anselm's formula, from chapter 2 of his *Proslogion*, 'that being than which no greater (containing more goodness) can be conceived'. Finally, God is simple in uniting in the divine being both causal and purposive explanations, since God is the necessarily existing cause of all, for the sake of goodness.

By both the main scientifically used criteria of probability and simplicity, God is the terminus of the quest for intelligibility and explanation in the universe. Nor is this just a matter of abstract intellectual investigation. To see that the universe is, both in its general structure and in its precise detail, a work of supreme wisdom, that it depends at every instant for its existence on a being of supreme goodness, and that the universe is destined to realise a goal of overwhelming goodness, is already to see the temporal world in a religious way. It can easily lead one to see the eternal in the temporal, to place the things of time in an eternal perspective. It can lead one to revere and contemplate with awe and admiration the one creator whose glory and power are seen in the works of creation, but whose being infinitely transcends that whole creation, and before whom the totality of space and time is, in Mother Julian's phrase, as it were a tiny hazelnut held in the palm of God. The search for wisdom, said Aristotle, begins in wonder. That search, for the religious thinker, ends in worship, as the mind is led from the contemplation of creation to stand, like Job, at last in awe before the immeasurable glory of the creator.

6

Breaking out of the Mechanistic Universe

God and the Laws of Nature

When Isaac Newton formulated the basic laws of mechanics, he was, as has been noted, motivated by the desire to trace the wisdom of a rational God in the works of creation. He assumed that God would not create a universe in which things occurred arbitrarily, or for no reason. There would be, he thought, a set of simple general laws governing the operations of physical substances. So events would stand in regular, predictable relations to one another, relations that could be described by measurable, quantifiable equations.

Newton's assumptions have been vindicated in a spectacular way by the success of science, which has shown how physical objects are indeed related in ways describable by elegant and mathematically simple expressions. Mathematics is the key to understanding the natural world, unlocking its nature as a set of complex outcomes of beautifully integrated fundamental laws. This is indeed a universe founded on intelligible wisdom, on that cosmic Logos that Christians see as the archetype of the whole creation.

Yet Newton's insight, perhaps the greatest intellectual revolution in human history, also led to some misleading and strictly unjustifiable inferences about the nature of the physical universe. These are best summed up in the famous dictum of the French physicist Laplace, that if he was given the initial state of the universe and all the laws of physics, he would be able to predict the whole course of its future development.

Thus arose the idea of the universe as a closed deterministic machine. Eighteenth-century deism built on that idea, to claim that a God of vast intelligence had indeed set the universe on its way, by determining its initial state and governing laws. But then God left it to its own devices, never interfering with the machine, which would proceed down predetermined paths to its predestined end.

There is a fundamental misconception here that the universe is capable of existing on its own, or that laws of nature are unbreakable absolutes that allow no exceptions or additions. If one really thinks that the universe is created by God, one is committed to the belief that not even one second of time can exist unless God actively sustains it in being. Laws cannot just be set up to operate on their own. What one would be saying is that God chooses always to go on creating the universe, from moment to moment, in accordance only with a few general mathematically elegant regularities, which God never varies.

Now that seems a rather odd thing for a God to do. It would limit the creativity of God enormously, and mean that God could never do anything that God had not decreed in setting up the general set of physical laws. There seems little reason for God to restrict the divine freedom to be radically creative in this way, even though there is very good reason to ensure that physical processes in the universe occur in accordance with intelligible principles.

What is needed here is a clear distinction between intelligibility and determinism. A process can be wholly intelligible without being deterministic. In fact, it may be a condition of the highest sort of intelligibility that processes are non-deterministic.

Determinism is the thesis that every event that comes into being is brought into being by some previous event, in such a way that there is no alternative to it. The cause sufficiently determines the effect. That is, given the cause, just this precise effect, and no other, has to follow.

Now it is true that if this was the case, one could explain the existence of every effect just by appealing to its cause. And one might trace the series of causes all the way back to a first, uncaused cause, which could be God. Some eighteenth-century philosophers, including most famously Immanuel Kant, were attracted to this thesis.

On the other hand, other philosophers, such as David Hume, raised the question of what it could be about a cause that *compelled* its effect to occur. Surely, Hume suggested, there is no logical necessity joining events in this way. So what sort of necessity could join physical events together, allowing no possible alternatives?

You could try saying that it is the laws of physics that make effects follow from their causes. But now the same problem recurs: what can compel effects to obey the laws of physics? Surely such laws are more like descriptions of what actually happens, not mysteriously existing principles that make things happen. It seems that there is nothing in the nature of a cause itself which could determine that one, and only one, effect should follow from it. In fact there is nothing about a cause, considered in itself, that could determine that anything at all should follow from it.

For theists, only God can compel things to happen, since only God has the power to give existence to finite entities. It is entirely up to God whether causes should have only one possible outcome. At this point, it might be said that since God determines every outcome, all events are determined after all. But this is not quite the same as determinism, which holds that each identical cause is necessarily followed by the same effect every time. If God determines all outcomes, identical causes could be followed by different effects on different occasions.

But what would be the point of that? I have already suggested that one important power God has is the power of free creativity. That is, God is not wholly determined, even by the divine nature. God can make radically new decisions, which do not follow necessarily from their antecedent conditions.

Appealing to Mozart, I tried to show how such creative acts could be intelligible, while not being necessary. God could act in various ways, but any way God acts will realise beauty and wisdom. The intelligibility of a Mozart symphony does not lie in the fact that every note has to follow from the one before it. It lies in the fact that, while realising many new and unexpected beauties, there are always patterns of complex order and subtle variations on basic themes, which cumulatively form an integrated, almost organic total structure.

The basic reason for this is that the highest form of intelligibility is a combination of both causal and purposive explanation. Events do need to connect with preceding events in ways that express continuity and pattern. That is the causal element. But at the highest level of personal existence and creativity, those past elements will be integrated in original ways, aiming at a goal of expressing beauty, of realising potentiality in a personal creative way. Creative imagination is the highest form of personal intelligibility. In its best expressions, as in a Mozart symphony, we see not only the growth of events out of past events – the causal dimension – but also the emergence of richer patterns which realise new forms of value – the purposive dimension.

Such a combination of causes and purposes provides a non-determinist view of intelligibility, and it can be seen at its best in personal or historical explanations. Suppose I see a person standing up, kneeling and bowing in repetitive but apparently pointless ways. If I want to make that person's acts intelligible, I need to provide some account of why someone is doing those things. I might try saying that the person is prone to some sort of obsessional neurosis – Freud might well have said that. Then I would need to delve into the individual's early childhood to see what terrible traumas had happened. A much better explanation, however, can be given by realising that the person is praying, following a ritual believed to express devotion to God. This is surely the right sort of explanation, without which one will always misunderstand such actions. There are causal explanations, too, of how the muscles work and so on. But what is needed for a correct account is reference to the beliefs and intentions of the agent. Since no purely deterministic physical account will ever mention beliefs and intentions, one can see that in many cases one can give a really intelligible explanation of a process, whether or not the process is deterministic, physically speaking.

Many philosophers would go much further – I would myself – and say that, in order to be intelligible, some processes actually rule out determinism. If human acts are really free, then it might be important that, in the very same physical situation, an agent can freely choose two or more different acts, and that choice is not inevitably caused by some existing physical state. Then one would have to say that the same cause (initial conditions) can be followed by different effects, if some agent really can act freely. Thus, when similar causes are followed by diverse effects, that is not necessarily because the process has become arbitrary or undirected. On the contrary, when those diverse effects are seen as parts of a free creative process, a developing pattern, a radically new expression of value, such a process has a greater intelligibility, a greater sense of purpose and directedness, than any process which is simply sufficiently determined, without any guiding purpose or value.

It is striking that the proponents of determinism did, in fact, come to lose any sense of purpose or value in natural processes. They came to see nature as a machine, working in accordance with pre-ordained, inflexible, impersonal rules. God was only needed as the first designer of the machine, and it was not long before such an unimaginative and rule-bound God could be pensioned off altogether. William Blake's painting *The Ancient of Days* portrays this mechanistic God, holding a pair of compasses in his hand, measuring out creation and forcing it to conform to his austere mathematical rules. Blake meant this to be seen as the anti-God, the death of the imaginative God, the death of a truly creative God, the harbinger of the death of creation itself.

If it is true that God is the imaginative creator of a patterned complexity that emerges from a fundamental simplicity and elegance of structure, then one can see how this universe is not likely to be a process that follows inflexibly from its initial state. It is fundamentally misleading to think of a whole complete set of laws of nature being mysteriously present at the Big Bang, waiting for particles to begin acting passively and obediently in accordance with them. The theory of cosmic evolution encourages us to think of the story of the expanding universe as the development of new and richer forms of beauty and wisdom, strictly unpredictable from their antecedents, but always remaining within the basic parameters of fundamental physical constants, general forms of patterned relationship which allow and even encourage the construction of more complex hierarchies of existence.

Freedom and Divine Causality

There is a further dimension to the evolutionary cosmic process. It is not only a product of the creative imagination of God. That might lead us to see all finite individuals as creatively chosen but still passive elements in the symphony of the universe, like the notes in a Mozart score, beautiful but inactive in themselves. We need to go back a step, and ask whether it is true that the creator determines all events to be exactly as they are. That might seem to be entailed by the thesis of creation, since nothing can exist without being given existence by God. But might God not create things that God does not wholly determine? The obvious case to consider is human freedom. God creates human beings in the divine image. If humans share in the imaginative creativity of God, they too will be partly undetermined, free to shape the future in patterns of order that arise from, but are not determined by, the past.

God could create beings with the power of imaginative creativity. God would hold them in being at every moment of their existence, but would not determine exactly what they would do. God would hold their creative power in existence, and would give them the power to make their own choices. God would have to co-operate with them by giving existence to the states they chose. So God might give existence to states which had not been chosen by God, but by finite creators, whose power is itself given by God. For example, God might give me the power either to play the violin or to go for a walk. The choice is mine, and whatever I choose, God is committed to creating whatever conditions are necessary for me to act on my choice. God must keep the violin in existence, however horrible the noise I make on it, and however much the neighbours complain. They cannot blame God, since God might have really wanted me to go for a walk, for the sake of my neighbours. They must blame me. In this way, many things might exist which God has not determined, though God must allow them to exist by some form of active co-operation.

It is a logical entailment of finite free creativity that many states exist that God does not intend. For anything that God intends must inevitably occur, and that would rule out finite freedom. God may *desire* creatures to act in certain ways, but strictly speaking God cannot *intend* that they do, if God gives them freedom. In fact, it is a consequence of the gift of freedom that there may be freely chosen states of which God disapproves, for free agents might have the power to bring about states of which God does not approve. God has given them that power, and might well have good reasons for not taking that power away, even if choices are made otherwise than God would wish.

Stress has been placed on the imaginative creativity of God, which brings about new values of beauty and wisdom. But Christian faith stresses even more the love of God, the decision of God to offer friendship to finite creatures. If the universe is such as to generate beings that are capable of friendship with God, we will have to reject decisively any idea of God as an absentee designer who leaves the universe to run on undeviating tramlines of causality. Nor is the Christian God a being who leaves finite agents free to make their decisions alone, without entering into active relationship with them. A God of love will ensure that the highest welfare of finite agents is found in their realisation of a conscious relation of love with God. God will make the divine being known to them, will guide them, invite their co-operation in shaping the future of the world, and will respond to their actions by aiding their good intentions and seeking to block the worst effects of their evil actions, so far as that is possible without destroying the causal structure of the cosmos and undermining creaturely freedom. Such a God will be in constant responsive interaction with the causal structure of the universe, an interaction that will increase as the universe generates conscious finite agents who are capable of personal relationship.

The reason deism is unacceptable to Christians is that it does not allow for personal interaction between God and human beings. In such interaction, God must be responsive to human thoughts and feelings, changing them by personal interaction with them. Otherwise the idea of personal relationship makes no sense. That again means that the universe cannot be regarded as a closed causal network. It must permit the emergence of beings that are in causal relationship with a being beyond the physical universe, the being of God. The physical universe, of which humans are an integral part, must therefore be constantly open to relationship with the creator. It must be an open system, not determined solely by laws of regular relationship between purely physical elements. Laplace must be wrong, and wrong in a very basic way.

7

God's Action in the Universe

The Open and Emergent Universe

A universe created by God must be open in a twofold way. Its laws must at a certain stage allow creative choice between open possibilities by conscious beings within the universe. That means that the basic laws of physics must allow for alternative futures, at least at a certain stage of complexity. The universe must also be open to causal influence from outside the physical realm altogether, influence from God. Such a universe would not be a wholly predetermined unrolling of a script already fully written at its beginning. It would be capable of unfolding in a quasi-organic way, in a number of diverse directions, and those directions must be causally influenced by the intentions of the creator. One might expect, then, that the universe would evolve in the direction of sentient life, but not down one inevitable path. Its processes would neither be wholly governed by necessity, nor by chance. There could be many critical points at which it could move in different particular directions.

At early stages of development, such moves might depend either upon God's creative decisions, or be left as indeterminate. There is reason to think that many of them must be indeterminate, since creaturely choices are made between open alternatives that exist in the system itself. There must therefore be a physical substratum of such choice, an openness within the system, which free choices can later utilise. Before rational choices come to exist in the system, such openness will appear as indeterminacy or randomness. However, physical processes will be stochastic (subject to precisely formulatable statistical laws, though not predictable in detail) rather than totally indeterminate, since there will be definite parameters of physical indeterminacy, which prevent nature becoming chaotic. One would expect that the structure of a theistic universe would be governed by stochastic but law-like intelligibility. That seems to be just what the situation is, in quantum physics. The behaviour of subatomic particles is subject to statistical laws which give a high degree of predictability in the overall process. We can, for instance, give the general half-life of atoms with a great degree of accuracy. Yet the exact behaviour of any electron is, according to Heisenberg's uncertainty principle, quite unpredictable in principle. So, at least at the subatomic level, the physical world has just about that degree of openness the theist would hope for.

At later stages in the cosmic story, selections made at open points within the general structure could largely depend on the choices of rational creatures. The theist would expect such choices not to be wholly autonomous, but to be governed largely by the relation of rational creatures to the creator and sustainer of the system. Finite agents might develop to become channels of divine creative imagination, co-creators of new forms of beauty and goodness. Or they might choose paths of self-concern and possessiveness which would obstruct or even corrupt the natural processes of creation. In such a 'fallen' universe, the creative power that should lead to greater co-operation and harmony of being would instead lead to conflict and mutual destruction. Such, unfortunately, appears to be the condition of our planet, whose natural processes have been diverted towards conflict and destruction by the cumulative choices of many generations of human and proto-human ancestors. This was always a possibility open to free agents with many competing desires, though it is a possibility God did not want humans to actualise. However, a God who willed to create these sorts of free agent could not foreclose the possibility of egoistic choice.

In such a world, the story of the interaction of God and rational creatures may well be the story of how God seeks, without undermining the freedom God has created, to turn humans back to the path of co-operation, harmony and union with divine creative power. It is in this general cosmic context that one should see the Christian claim that God has acted in the historical person of Jesus, and in the cluster of crucial historical events around the life of Jesus, to show how God seeks to unite and will unite all assenting persons to the divine, without undermining human freedom. The Christian gospel, or 'good news', is that in Jesus God the creator acts to provide a paradigm case of the way in which God desires to reconcile a corrupted world to its creator, and to restore the original divine purpose for the earth. He is the embodiment or incarnation of the Christ, the liberator and ruler of the created order, for this planet.

The universe would not, on this account, be sufficiently determined, but neither would it move in a wholly random or chaotic way. Rather, it would be drawn towards the development of forms of life which could come to know and love God, by an intelligible but free unfolding of its potential being. It would provide those forms of life with an environment in which they could take control of their own futures and the future of their planet. And it would be open to the persuasive and co-operative power of the love of God, seeking to restore and complete the original purpose of generating a community in which new values could be continually created, appreciated and shared. Through such a community the material universe could become aware of its own created nature and capable of creatively expressing its potentialities for good, in conscious co-operation with its creator and sustainer.

This theistic account does seem to fit the nature of the physical universe and its laws, as we observe them. The Laplacean story belongs in a pre-Rutherford universe, when it was thought that atoms were indivisible particles with a finite set of measurable properties. In such a 'billiard-ball' view of the universe, the atoms go on banging into each other, forming complex patterns and structures, no doubt, but never amounting to more than a moving arrangement of balls on a finite table.

Since Rutherford split the atom in 1919, this sort of closed determinism has become less and less convincing. It always was an expression of sheer faith, a rationalistic faith that the universe simply must be deterministic, or it would be chaotic and inexplicable. That account of intelligibility, I have suggested, ignored creativity, contingency, purpose and personhood completely. It was the result of the mistaken idea that the physical universe was a closed system, in which everything that ever happened was predictable from a first simple set of axioms.

It is now generally accepted that there is no set of axioms known to science from which every particular occurrence in the history of the universe can be deduced. It is almost certainly not possible even to deduce the laws of biology, psychology and neurophysiology from some set of laws of physics. Most scientists now have a picture of a hierarchy of contingent laws, realising sets of rule-governed events that are not predictable solely from laws that govern earlier stages of the cosmic process, lower in the hierarchy of complexity. The structure of the material universe seems to be such that it realises successively more complex sets of rule-governed interactions, each level cumulatively building on the levels below.

The clearest case in which higher levels of complexity seem non-deducible from lower levels is that of consciousness. Some philosophers make heroic attempts to argue that conscious events at the highest level are actually nothing but very complex arrangements of low-level particles. However sophisticated these attempts get, they can never evade the obvious fact that the phenomenal properties of consciousness (like 'seeing something green') are different in kind from the properties of quarks or electrons (like electromagnetic charge, spin or 'charm').

We can be more certain that 'greenness' is not a complex combination of positive charge and spin than we can be of any sophisticated argument that one is reducible to the other. This seems a very clear case in which organised complexity at one level, the level of the brain, facilitates but does not entail the occurrence of events at a higher level, the occurrence of phenomenal (perceived) properties. After all, the same brain events could give rise to different phenomenal properties. We just have to look and see what correlations do, in fact, exist. We need not doubt that there are such general correlations. The occurrence of phenomenal properties is not random, but law-like and regular. Moreover, they build on brain events, adding new principles of correlation which do not contradict those of lower levels, but which utilise them as a basis for building new forms of law-like relationship.

In the case of consciousness, there is little doubt that new properties come into being, which have never before existed in the physical universe. When Einstein had the thought, '$E = mc^2$', that thought had, so far as we know, never existed before, certainly not on the planet earth. It was an absolutely new thought, which no amount of knowledge of the position of electrons in Einstein's brain could have predicted. The thought depended in some way on the complex structure of Einstein's brain, and therefore on the cumulative organisation of the organic, molecular, atomic and subatomic levels that caused that brain. But while the thought emerged from Einstein's brain, it was something quite different from all the movements of molecules of which the physical brain consists. Thought, it seems fairly clear to most people, is an emergent, new property.

If there are emergent properties in an open universe, one might well think of God as exercising a continuing influence on the developing complexity of the cosmos, ensuring that conscious organic beings do come into existence, though perhaps not determining exactly what they will be like. God will not simply set the universe going at the first moment of its existence. Rather, God will be continuously involved both in sustaining the universe in existence at every moment and in actively guiding the way it develops, within the limits of the general laws of its intelligible structure.

The God of the Gaps and the Universal Dictatorship of Law

Talk about God as a causal influence in the emergent processes of the universe may be dismissed as an appeal to a 'God of the gaps', a God called in to explain gaps in scientific theories, which will inevitably one day be filled by a more complete theory, thus eliminating the need for such a God. But those who dismiss it in this way are in fact relying on a dogma of their own, the dogma that scientific theories can explain everything. That is, there is some set of general law-like statements in accordance with which absolutely everything in the universe happens. But that is precisely the view of closed determinism, on which contemporary physics throws so much doubt.

Closed determinism operates with a 'one-level' view of the physical universe, and it does not fit an emergent universe with different and irreducible (non-deducible) levels of complexity, order and relationship. Nevertheless, there certainly are laws of physics. Even if they are selections out of wider possible ranges of laws, they do apparently disallow non-law-like behaviour. So a sort of determinism could still be true, whereby, at a given level, all events are subject to general unbreakable laws. The equations of physical theory are, after all, usually deterministic in form. If you put in precise values for variables in one part of the equation, the solution follows invariably. This does suggest a sort of determinism in nature. It could be just a messy complication that new laws seem to be introduced from time to time in the history of the universe, when new levels of complexity emerge. There is after all, it may be said, a universal dictatorship of law, from which nothing can ever escape. Non-law-like events, would be 'gaps' in the system, and they cannot be allowed.

If the equations of physical theory are deterministic, the question to ask is whether those equations exactly correspond to actual events in the physical universe. Most contemporary scientists would be very reluctant to assert such an exact correspondence. Laws are seen as ideal and abstract models, more like maps of a landscape than photographs. A law governing the motion of molecules of gas in a confined space, for example (Boyle's law), states how molecules behave in a system isolated from all outside influences (where the temperature is kept constant), where one can be reasonably sure that one has considered all the relevant causal factors, and where one can measure the values one is interested in with precision.

In the confined space of a jar in a laboratory, one can be fairly sure that the pressure of the gas will bear an invariant relation to its volume, that no other factors will interfere, and that one can measure both pressure and volume accurately enough to give a highly reliable prediction. Technically, however, no physical system is ever completely isolated. In the strange world of quantum physics, Bell's theorem states that subatomic particles light years apart will remain causally entangled if they have once interacted. So events on the other side of the universe have some effect on events on earth. Fortunately, we can ignore them for most practical purposes.

However that is not always so. The development of chaos theory in recent years shows that very small differences in initial conditions, or tiny influences occurring at critical points in a dynamic process, can cause large-scale and wholly unpredictable consequences in macrocosmic systems. The planetary weather system, for example, is just such a dynamic system, and tiny changes in one place can destabilise the whole system, producing huge and unpredictable effects elsewhere in the system. Weather forecasters have a perfect excuse for why they can never predict the weather accurately over longish periods of time. It is not their fault, it is the nature of the system. Chaos theory is itself deterministic in the way its differential equations operate, but it shows that the tiny influences we usually ignore can, in the right conditions, have major effects, of which our predictive models can take no account. This, in turn, shows that the laws we are able to formulate will never exactly correspond to reality. They will never be able to take into account the total interconnectedness of nature, which introduces causal influences into the system that we cannot discover. Such influences render the system unpredictable in the long term and as a whole.

At that point, it is theoretically clear that we cannot rule out the existence of causal influences that are undetectable by us, and which may be non-computable. We can never be sure that we have specified all causally relevant properties exhaustively. That is why scientific laws always contain a silent 'other things being equal' clause. The volume and pressure of a gas are invariantly related, other things being equal. But sudden changes in the gravitational field, for instance, may introduce fluctuations that we will choose to ignore.

In addition, in choosing just those two properties of gases, we are excluding many other properties that we are not interested in at the time. But those properties, like the charge and spin of the electrons involved, are parts of the system. We can place no theoretical limit on the number of properties that may be causally relevant to the system. Some of them, like the 'dark matter' of the universe, may be inaccessible to us. In pointing to a ratio that is invariant in normal conditions, we are a very long way from proving that every event in the universe follows necessarily from its preceding physical state, in accordance with some general law.

Finally, we now know that we cannot measure all properties with absolute precision. There are limits to the accuracy of any measurements we can make. Heisenberg's principle of uncertainty points to one of these limits, that it is not possible for us to measure accurately both the position and momentum of a subatomic particle. Even if there are 'hidden variables' which actually determine how things are, the way in which observation interferes with a physical system makes it impossible for us to observe them. Once again, it is impossible to say whether there is a fundamental indeterminacy in the basic laws of physics – as most physicists believe – or whether it is just that we can never formulate totally accurate physical laws that will fit the real world with perfect correspondence.

In an important sense, the equations of physical science are just too precise to get an exact fit with the fuzzy world that they attempt to map. Their precision is bought at the price of abstraction, of isolating a set of variables between which we can establish regularities that hold for most normal purposes. This fact does not license a commitment to determinism; it even suggests that there may well be causal factors at work in the physical universe which cannot be captured by the deterministic equations of physical theory. The dictatorship of law is not so universal after all.

This does not mean that there really are laws of physics that are 'broken' from time to time. It means that the laws of physics we are able to formulate do not, and cannot ever, provide us with a totally comprehensive, exhaustive and accurate picture of the real physical world. Actual events are parts of a total system of interactions, some non-measurable, some inaccessible, and we can never state all causal factors that are operative at any time. If this is so, we do not have the problem that God is excluded from a closed deterministic physical system, and has to interfere in it to perform specific actions. Rather, God may exercise causal influence on a system that is essentially open to the influence of purposive causality both within the system (in humans, for example) and from without.

If the universe is an open, emergent and interconnected system, and scientific laws are ideal models for understanding regular and quantifiable connections within it, there will always be some features of the physical universe that laws of nature cannot capture. In the old Laplacean picture these would be thought of as gaps in an otherwise seamless web of deterministic causality. But in the new physics there is no deterministic web, with gaps which will one day be completely filled. There is a hierarchy of dynamic and interconnected systems, regular and quantifiable parts of which can be modelled in mathematical terms. But there is much that cannot be so modelled. One would expect that personal acts, which express unique creative intentions, would be subject only to very approximate modelling, at best. Such acts are not gaps in a mechanism. They are precisely those unique and creative features that characterise the most complex physical systems, and disclose the character of the material cosmos as not a blindly operating machine, but an environment exquisitely formed for the development of personal consciousness and relationship.

Human and Divine Action

When we go on from considering the purely physical nature of the cosmos, and begin to think about conscious beings and their actions, we get a much fuller picture of how a creator God relates to the universe. We can begin to redraw the restrictive picture of the world as a closed and mechanical system. When human beings act intentionally, for instance, it is their thoughts and desires that account for what they do. But there is no *law* that, if I have a certain thought, I will do a certain thing. For a start, there is no way of measuring thoughts, so I cannot set up one of the equations that are essential to scientific laws. Then, my thoughts are not regularly connected with my actions in invariable ways. Humans remain largely unpredictable. Thoughts influence behaviour, they make things happen for a purpose or reason, but they do not follow some general and universal law. The law-like explanations that are so helpful in physics and chemistry will probably not work for human beings, and they will certainly not work for God's actions.

If I am really a free agent at a specific time, that means that there are a number of different things I might do at that time. I can now, for instance, carry on writing or have a cup of coffee. Both those alternatives are open to me, and no law of nature determines which I am going to do. If this is true, the physical universe must be such that it often places open alternatives before personal agents. There have to be alternative futures. That means that the laws of physics cannot wholly determine the future in every respect. They must be to some extent non-deterministic. At the quantum level, it seems that indeterminacy does exist, but there is not a great deal of significance in the undetermined, but probabilistic, movement of electrons from one energy-state to another. Nevertheless, in complex and finely integrated systems, in the human brain for example, such minute state-changes may have macrocosmic effects which are significant. They can change the parameters of a physical system, so that it develops in quite a different way than it would otherwise have done.

Generalising from this result, it seems that a medium-scale physical system may be sensitive to influences that will change the parameters of the system – not just quantum 'jumps', but perhaps other more direct influences. One example might be the way in which, in embryogenesis, the development of cells to have specific functions in the body is influenced, not by anything internal to the cell itself (since all cells in a growing organism start off as identical), but by the development of the organic system as a whole, which switches on different genes (pieces of DNA), according to their place in the organism.

Quantum influences work 'from below', changing the substratum of the system. Organic influences work 'from above', changing the boundary conditions that interface with parts of the system. In a third type of influence, the human mind, as the highest level of a complex physical system, is able to influence the system in response to stimuli received from that system. At this level of complexity, the system itself becomes able to select between its alternative futures, in accordance with stimuli received and the formation of conscious intentions. The whole physical world takes different paths, depending on the consciously formed intentions of human minds. The intrinsic openness of physical systems, which remains largely implicit or not consciously directed ('random') up to that stage, now becomes an important feature of physical reality.

If there is a creator, it is probable that God would make things happen for a purpose, and would often act in particular ways to effect such purposes. If God's purpose is to realise a relationship of love with finite persons, God's particular actions will become more frequent and interactive when such persons exist and develop the capacity to become aware of and responsive to God. This will be a fourth kind of causal influence on physical systems. It will be strictly 'supernatural', since it is exerted from outside the physical system itself. But it would be highly misleading to regard it as an 'interference' with the system, since, from a theistic point of view, the whole system has been developed precisely to make such interaction possible. It is the goal of the physical cosmos to enter into a conscious and active relationship with God, its spiritual foundation.

We might say that the whole of the physical cosmos is oriented to transcendence, to that which lies beyond it but is the only source of its reality. It is not a self-enclosed system, but a developing system which strives to a goal beyond itself, in relation to which it can realise its deepest potentialities. If we think of relationship with God as the goal of evolution, then when God acts to establish such relationship, it will be misguided to call that an interference. It will be the realisation of the goal, or at least one expression of the goal, of the cosmic process.

In the vast times and spaces of the universe that contain no personal forms of life, the creative acts of God will almost certainly not be interactive and personal in this way. God may, like John Conway and the Cambridge mathematicians who designed the 'Life Game', enjoy setting up seemingly simple but amazingly fruitful laws that will produce endlessly interesting and original patterns of beauty and order. The devising of such laws requires great intellectual application, and the appreciation of their consequences requires great sensitivity to beauty. So, while one could put the existence of such laws down to chance, it is entirely plausible to attribute their existence to an intelligent ordering for the sake of beauty. Such divine action will not be some sort of addition to the ordinary laws of nature. It will be precisely a selection of a highly ordered set of such laws, and the ordering of matter to change in accordance with them.

In other words, through millennia of cosmic evolution, God's actions may well consist, not in particular local interactions with physical phenomena (like human actions), but in the generation of new patterns of law-like order, which build on established patterns and add new general principles of cumulative organisation. God will not act to interfere with laws of nature. God will act precisely to devise and implement laws of nature which generate new patterns of complex beauty and order. But in doing that, God will ensure that such laws will facilitate the eventual realisation of the purpose of the system – to generate persons. And God will ensure that the cosmos retains the right balance of openness and intelligibility to form an appropriate environment for personal life, when it emerges.

Even when fully personal life has come to exist, in most cases divine action will be a causal influence that works within the parameters of probabilistic physical law. God might console me when I am in trouble, or might heal someone in answer to my prayer. This can happen without breaking any laws of nature, just by using the possibilities that exist in the system, which might be influenced for good by my relationship with God. But there can be exceptional cases, particularly at later stages of the process, when conscious life has evolved, when the regular laws of nature are suspended. This happens in the case of miracles, widely attributed to the intercession of prophets and saints, when events occur that are beyond ordinary scientific explanation. Even in these cases we should think of God's action, not as an arbitrary interference, but as the raising of material nature beyond its purely natural, autonomous, powers, to attain its destined purpose by mediating the divine presence and purpose. So the healing miracles of Jesus show the presence and power of God in him, a power that makes others whole and gives true health to human lives.

In other words, supernatural action completes and fulfils the natural. It does not contradict the natural, but raises it to its proper form of being. God's actions in the universe before conscious beings existed would probably have consisted in the selection of laws that would ensure that a realm of personal agents came to exist by a gradual and emergent process. God's action in relation to human persons is intended to raise them to fully conscious relationship with God, to be channels of divine wisdom, power and love, to be co-creators of value and guardians of God's creation.

8

Creation, Suffering and the Divine Purpose

The Autonomy of Nature

The whole universe is in some sense the act of God, or a continuous series of divine acts, so one might think that God must intend everything that happens in the universe, willing each particular state to be exactly what it is. But the universe is not such that at every time it directly expresses some divine intention. It is not possible, for example, to believe that God would directly intend the suffering of finite creatures, or the actions of creatures that break those moral laws that God lays down. God cannot intend that one person murders another, for God expressly forbids such actions, and it is incoherent to suppose that God intends the very same thing that God forbids.

This universe is one in which much suffering exists, and many evil actions, harming other sentient beings and destroying natural processes, are performed. If we believe that God is the creator of the whole universe, then God must create those things. Yet if we believe that God is good, then God cannot *intend* those things. We are forced up against a possibility that many traditional religious believers have found hard to accept. God, the one and only source of all beings, must be the source of beings that God does not intend.

The Bible does not shrink from this thought: 'I form light and create darkness, I make weal and create woe': so writes the prophet Isaiah of the creator (Isaiah 45: 7). Can we really think that God intends to create affliction and suffering? We have already seen that the divine nature is necessary. It cannot be other than it is. There is no alternative to it. It contains all possibilities in itself, and is not free to change those possibilities. Now, we do not know what the limits of necessity in God are. We do not know what possibilities are so linked to other possibilities that one could not be realised without the other. It may be, then, that if God intends to realise some possible states, they are necessarily linked to the realisation of other possible states that God does not intend.

Suppose that God intends to create beings that are free to determine their own futures. The existence of such freedom may entail the possibility that destructive and harmful choices could be made. That possibility could not be eliminated without eliminating freedom itself. Those possibilities are necessarily linked together. That does not mean that evil will exist. But it does mean that it can exist, if free agents choose to realise it, and God cannot prevent it existing as long as God wills agents to be free.

According to Christian belief, this is a planet on which human beings have willed destruction and evil, against the wishes and purposes of God. God intended that humans should be free and should freely grow in the love and wisdom that God could give them. But humans chose otherwise, they chose to reject the divine purpose, and they realised evil. God does not intend that evil, but cannot prevent it, as long as God continues to will freedom. This is one way in which God may create and sustain states of being that God does not intend. They are unpreventable by God, as long as God continues to intend certain other states – namely, the existence of free finite agents. That is one way in which God can be the only source of states that God does not intend.

There may be other necessary connections between possible states, to which human knowledge has no access. It seems probable, for example, that a universe that is truly emergent is one in which some measure of conflict and suffering will necessarily exist. Old forms have to die away, to make room for new. And it may be partly through competition and conflict that new forms come into existence. In this way, the distinctive values that only an emergent, evolutionary universe can realise – values of courage, tenacity, creative adventure, as well as values of compassion, co-operation and self-sacrifice – will not be able to exist without the existence of some sorts of suffering that God does not directly intend.

We might say that God intends the values, the goods, that only such a process can realise. Therefore, God does generate the whole process intentionally. Yet God does not intend the suffering and conflict that the process entails, or at least makes unpreventable by God. They are foreseeable and inevitable consequences of God's intending these values. And it may be that any world of rich and complex values that God can create must contain some such unintended consequences, because of features of the divine being itself that even God is not free to change.

So God may create this universe for the distinctive goods it contains, and do so intentionally. But there may be many states in this universe that God does not intend, many states that God forbids (but does not prevent), and many states that God actively opposes. The reason God cannot actively and unilaterally eliminate such states is that they – or states very like them – are necessarily implied as possibilities in a universe like this. This universe contains great goods that could not otherwise exist at all. Even God cannot intend a universe like this to exist, and at the same time eliminate all evils from it.

Divine Necessity and Freedom

At this point we are faced with what most people feel is the greatest problem confronting anyone who believes that God created this universe. It may all be elegant, mathematically beautiful, amazingly intricate and awe-inspiring. But it seems to contain so much cruelty, misery and suffering that one can wholly sympathise with those who cannot accept that the creator is good. To many the creator of this universe seems cruel, or at best indifferent. Does modern science finally demolish the idea of a good God who cares for individuals? Or does it, on the contrary, give some new insights, at least at a theoretical level, into the goodness of the creator?

A large part of the problem, I think, lies in what has been called the 'fallacy of omnipotence'. We think of God as having absolutely unlimited power, so that God could create any logically possible world at all, any world that we could describe without apparent self-contradiction. So God could create a much better universe than this, perhaps with no suffering in it at all, or God could create this universe with much less suffering. If that is the idea of God's power that we have, then it is impossible to justify the existence of so much suffering as there is, when it does not have to be here.

The scientific understanding of the universe suggests that this idea of omnipotence is a fallacy, a mistake, though it is an easy one to make if one wants God to be as powerful as possible. It is a mistake, firstly, because the scientific quest for explanation suggests that the 'absolute explanation' of the universe has to lie in a necessary being, a being that does not choose its own nature. We have seen that God does not choose the array of possibles. They exist by necessity, as parts of the divine nature. God must choose from among a set of necessarily given possibles. In this sense, God's choice is not entirely free. Maybe God has to create some universe, just as God has to exist, and maybe any universe God creates has to have a certain structure, which is laid down unchangeably in the being of God itself. So we should not think of God as able to intend absolutely anything at all. Maybe God has to create a universe of a certain sort, because that is part of what God essentially is.

There is a second reason for thinking that God cannot do absolutely anything, which is suggested by the scientific worldview. Cosmologists often stress how everything in this universe is strongly interconnected, so that the fundamental physical constants and laws need to be exactly what they are, if there are to be conscious life-forms like human beings in the universe. We cannot change those basic constants and leave everything else just the same. So we cannot take away central nervous systems, with all their potentiality for causing pain, and leave the rest of the universe just as it is.

We still do not understand this interconnectedness of the universe very well. But it has become entirely plausible to say that beings like us simply could not exist in a universe with very different laws. Some physicists even claim that the laws of this universe are the only laws that could produce any life-forms at all, but that seems to me to go far beyond what we could prove. Still, the point would be that, if God wanted to create human beings, the fundamental laws of the universe would have to be very much as they are, and they would, it seems, necessarily involve all the possibilities of suffering that we see. If God wants to create life-forms very like human beings, God will have to create a universe with all the properties this universe has. To put it bluntly, God could not create *us* in a better universe, or a universe with fewer possibilities of suffering in it. God's choice is not between creating us in this universe or in a better one. We would not exist in a better universe. So God's choice is between creating us, in this universe, and creating different types of beings altogether.

Science can give us the idea of a God who necessarily creates some universe, one of many possible universes whose structures are each governed by necessary interconnections that make the values of that universe inseparable from at least the possibility of suffering, and from a good deal of actual suffering. That idea is very different from the idea of a God who can do absolutely anything, and who could therefore get rid of all suffering quite easily and leave everything else the same. In fact, however, it is much nearer to the classical idea of God in Jewish, Christian and Muslim traditions than is the God of arbitrary unlimited power, who has probably only existed in regrettable popularisations of religion. The biblical book the Lamentations of Jeremiah puts the point clearly: 'The Lord . . .

though he causes grief, will have compassion, according to the abundance of his steadfast love, for he does not willingly afflict or grieve the sons of men' (Lamentations 3: 32–33). God is the cause of suffering, but does not intend that suffering just out of maliciousness or perverse pleasure. God will ensure that the overwhelming good, of which it is a condition, will at last be realised.

The Goodness of God

We might be able to understand such an idea of God. We might stand in awe of God as the ultimate source of all beings, and admire the immensity and intelligible beauty of creation. But would we call such a God 'good', and does such a God care for individuals? At this point we have to leave the world of science, and turn to the experiences of the great saints and mystics of the world's religions. The first thing they have to tell us is that in contemplative prayer it is possible to have an intuitive knowledge of the supreme goodness of God. It is very important to see what is meant by 'goodness' in this context. The divine reality is good, in that it is a reality of pure beauty and bliss. If 'the supreme good' is an object of contemplation which is wholly fulfilling and satisfying, then the infinite beauty and bliss of the divine being, as experienced by the mystical consciousness, is the supreme good. It is an experience that outweighs every other in value, and which is supremely worth pursuing. Plato, in his dialogue *The Symposium*, expresses it well:

> This is the right way of being initiated into the mysteries of love, to begin with examples of beauty in this world and use them as steps to ascend continually to absolute beauty as one's final aim . . . This above all others is the region where a truly human life should be spent, in the contemplation of absolute beauty . . . One who contemplates absolute beauty and is in constant union with it . . . will be able to bring forth not mere reflected images of goodness but true goodness, because one will be in contact not with a reflection but with the truth.

In Plato, a connection is made between the contemplation of beauty, human goodness and love. One who contemplates beauty can rightly be said to love that which is most worth loving. And, since Plato holds that one becomes like what one contemplates, the contemplator of beauty is one whose life becomes beautiful, an embodiment of that true goodness which lies in the possession of what is most worthwhile. In this sense, God is supreme goodness. To love God is the highest goal of human life. To attain God is to have one's life enfolded in perfect beauty, and to become an image of goodness to others. That is the primary sense of divine goodness, for every traditional religion.

The Creativity of God

Thus we can move from awe and admiration of God to love of God. But what reason is there to think that God loves and cares for individuals, in this world of so much pain? It is tempting to think, as some followers of Plato did, that this universe necessarily 'overflows' from God, and has to be the way it is. But the universe is not really good, God does not positively intend it to exist, and the best thing is to escape from worldly life into the pure contemplation of God. Christians, however, have always stressed that the created universe is good, even though it has been corrupted by the selfish desires of creatures.

It is good because it is a place where lots of good things exist, which otherwise would have had no existence at all. They can be very simple things, and no less valuable for that. Things like enjoying a beautiful view, drinking a pint of beer, skiing down a mountain slope, or making love – these are good things, which need a material world and embodied beings if they are to exist. So there is a very good reason for God to create some universe, which is that it will realise goods that otherwise would not exist. Moreover, each possible universe has its own unique good, which can only exist in that particular universe, or one very like it.

Created goods are not just enjoyed by creatures, which otherwise would not exist. They actually add something to the being of God itself. When God creates, God expresses the divine nature in a way that would otherwise have remained only potential. And God, too, can experience the unique goods that otherwise even God would not have experienced. We might say that creating universes changes God, or at least that it expresses the divine nature in a quite distinctive way.

One of the most decisive intellectual transitions from the ancient and medieval worlds to the modern world is the fading of the Platonic vision that the timeless realm of Forms is more real than its faint images in space and time. For theologians like Augustine, Anselm and Aquinas, the world as it is conceived in the mind of God is more real than the actual space–time world, which suffers all the deficiencies of transience and finitude. For most people born in post-sixteenth-century Europe and America, however, it is the space–time world of particulars that is real. The world of Forms is at best an abstraction, and worlds conceived by God are merely possible, without having full actuality.

For early and medieval Christian theologians, God could be thought of as conceiving all possible worlds in the divine mind, and as enjoying their full beauty and intelligibility, without actually having to create them. For most of us, the contemplation of possibilities is less satisfying than the contemplation of actualities. So if God is to enjoy many finite forms of beauty, they actually have to be created. The God of Aristotle could rest content in the contemplation of God's own reality. But then that God was not, and could not ever be, a creator. If creativity is an essential characteristic of God, then there will have to be an actual world in which that creativity is exercised and enjoyed. The creation of some finite world will be an essential characteristic of a God who naturally seeks to express the divine being in a creative and imaginative way, so as to realise and enjoy endless forms of beauty and goodness.

The Love of God

In addition to this consideration that creativity is part of the divine perfection, Christian faith makes central to its vision the claim that 'God is love' (1 John 4: 8). I think what is intended here is that God seeks to express the divine being as one that realises its nature in relationship to others in appreciation, co-operation and compassion. Love essentially goes out from itself to another, both in sharing friendship, in seeking to bring the other to its proper perfection, and, not least, in receiving from that other the distinctive values it creates.

Many ancient and medieval theologians held that such love could be completely expressed within the divine life of the Trinity. As they saw it, Father, Son and Spirit could be bound together, giving, receiving and sharing love in an indivisible communion of being. Yet it would still be true that such love was confined within the divine being. It would really be the love of God for God, not for another, and it might be doubted whether even such a sophisticated sort of self-love is what is depicted by the self-sacrificial love of Christ on the cross, giving his life for others, that they might live.

The symbol of the cross suggests that the creation of finite persons involves a definite risk for God, the risk of rejection and suffering. The creation of others, who would be free to receive or reject God's love, whom God could both delight in and feel compassion for, would make possible a real community of being between God and that which is not God. If God really is love, it will be natural, though perhaps not inevitable, that God will create others who are free in relation to God, giving them autonomy and control over their personal lives, but also inviting them into friendship and a sharing in the divine life, which would be their deepest fulfilment.

The risk of such creation would be that creatures would fail to enter into loving relation with the creator, and fall into a self-chosen estrangement from God and eventually from love itself. If God is a God of creative love, as the Christian doctrine of the Trinity suggests, then God might essentially create some universe of sentient beings in which love can find expression, even at the risk of rejection and at the cost of suffering even to the creator. The symbol of the cross seems to show, in a way that some theists find shocking and even offensive, that God is rejected and made to suffer by the very beings that God has created. Even in God, the cross seems to say, there are states that God would not intend or wish to exist, but that God permits and accepts as part of the price of creating a universe like this.

We might therefore say, largely on the basis of the distinctive revelation of the nature of God in Jesus, that the creator is a loving God. For the creator intends that there should be a community of loving persons (perhaps many such communities) that will find their fulfilment in a common relation to God. But the price of creating a universe in which such a freely chosen community can come to exist is that pain and suffering will result when persons hate and try to destroy each other. In such a universe, a loving creator will not be able simply to eliminate all suffering. But it may be possible for God to share in that suffering, to redeem it by bringing some otherwise unobtainable good out of it and, in the end, to ensure that the ultimate triumph of the divine purpose of love will be achieved for all who do not explicitly reject it.

The Hope of Eternal Life

Belief in the ultimate triumph of God's purpose is perhaps the most important element of all for explaining how Christians can believe that God cares for every individual, when so many suffer and die tragically. The scientific understanding of the universe may help us to see how human life is emergent from a simpler material order, and must carry the marks of its long evolutionary history with it. In this order, conflict and actual suffering are inevitable, though in forms and degrees that have been made incomparably worse by human sin. The existence of human agents is only possible in a universe with some actual suffering, because of the conflictual processes of evolution. It is only possible where there is a great deal of possible and divinely unpreventable suffering, that can be brought about by evil choices.

Nevertheless, if theism is true, the goods this universe will realise are so great and their relation to the actual and possible suffering they involve is such, that this universe is overwhelmingly worth creating. It is much better that it exists than not, and it must be possible for every created being to come to a stage at which they will agree with that, if they see things as they truly are. That means, I think, that an essential part of the Christian good news is that this earthly life is not the end of individual existence. All reflective sentient beings are offered a life of endless joy, which they could not have obtained without first having been born on earth, and which will far outweigh the sufferings they experience on earth.

That is the decisive factor which makes it possible for Christians to say that God cares for every created individual. The natural sciences do not seem to have any information to give us about an afterlife. They can take us as far as seeing that this universe, with all its suffering, may be somehow necessary to our existence as human beings. But they cannot show that God is a being of supreme goodness, who wishes to enter into loving relationship with communities of created persons, and who will ensure that all such persons will experience eternal happiness, unless they explicitly reject it. It is saints and mystics who claim to show us such things. For Christians it is Jesus of Nazareth who by his life and teaching shows the love of God, who by his passion and death demonstrates that God shares in human suffering, and who by his resurrection promises eternal life. It is hardly surprising that revelation tells us things about God that science does not know. It is more surprising, at first, that science tells us things that help a little in understanding the deeper purposes of God. But then, on reflection, it is very natural that as we understand more about the created universe, we will learn something more about the nature of the God who created it.

Constraints on Divine Action

I suggest rejecting the idea that God can do absolutely anything at all. But we can still call God 'omnipotent', because God is the only source of all other beings, and because there is no other possible being that has greater power than God. We may be sure that such a God will not only create a universe for the sake of the goods it contains, but will also act within the created universe to ensure the eventual fulfilment of the divine purpose. However, the nature of any universe that God wills is such that there will be definite constraints on the actions of a God who acts through the persuasion of love rather than through irresistible power. I have suggested that this universe is ordered towards the existence of communities of rational beings who can enter into personal relationship with God. It must therefore be 'open' in structure, to allow both human choices and divine acts of disclosure and response that will have a free and spontaneous nature. It must also be law-like in its basic organisation, so that rational beings can come to predict and control their world, to a large extent. It must be emergent, becoming cumulatively more complex in organisation, if matter itself is to develop into a consciously directed sacrament of spirit. It must consist of a set of interconnected and finely tuned systems, if it is eventually to form one organic unity capable of mediating the divine life.

These structural features of the universe will not just come into existence with the evolution of human life, as though all the fundamental laws of physics would change at that point. They must be characteristic of the universe from its origin, and they will determine the sort of universe this is, into which humans will be born as integral parts, not alien intruders. It follows from all this that the cosmos must have a certain independence from God. It cannot be directly God-determined in every respect.

Where there are laws, many particular states must follow from the operation of those laws, which could not be intended if they were considered as isolated states. For instance, God would never intend that a person should be crushed to death by a large rock. But if volcanic eruptions happen in accordance with the general laws of geophysics, which God intends to exist, some people may be crushed by large rocks. God cannot retain the laws and preserve all people from harm, except by a continuous series of miracles – and that would undermine the structure of law.

Where there are open, undetermined futures, God cannot step in to determine them. So in stochastic, probabilistic systems, many things will happen 'by chance' which are harmful. If the mutations that drive evolution are probabilistic, many of them will be harmful, and only some will be helpful. That is the necessary price of building indeterminism into a physical system.

Where physical systems are cumulatively complex, each step of the process must build on previous steps and be constrained by structures that are left over from past history. Humans, for instance, have genetically programmed dispositions to lust and aggression that were naturally selected over millions of years of evolutionary competition. Although they were once beneficial, they now appear as unfortunate constraints on acceptable behaviour, and they cannot simply be eliminated. There are ways of building constructively on them, but the success of such strategies will largely depend on new decisions and attitudes, which cannot be compelled or predicted. So one would expect the developmental process to be ambiguous and halting, rather than maximally beneficial at every stage. That is the price of creating a truly emergent universe.

Where physical systems are bound together holistically, each part will be sensitive to changes in other parts. What is beneficial to the whole will sometimes be harmful to particular parts, and competition, as well as co-operation, will be an inevitable feature of any quasi-organic system. To take a very mundane example, if you want a beautiful garden, as opposed to a chaos of weeds, many plants will have to be rooted out, and others will have to be severely pruned. The good of the whole, we might say, requires harming some of the parts.

In the physical world, it will not, in general, be possible to adjust one part to improve it, without having effects throughout the system that may not be beneficial at all. Physicists believe that the influence of the gravitational pull of any body extends infinitely throughout space. So changing any physical body will produce some changes, even if minute, everywhere in the universe. Sometimes, at critical points in the system, even very minute changes can have large-scale effects. Thus, God will not be able to consider the welfare of particular parts of the universe in isolation from the total system. That will introduce a major constraint on the possibilities for divine action in such a universe.

Divine Action as Co-operative Influence

How might one expect God to act in such a universe, before communities of rational beings have come to exist? Complete divine determination of all things is ruled out. It might seem that God would set up the system and leave it to run itself, only coming into action when sentient beings are able to relate to God. This picture of God setting up an autonomous system, and then observing what happens to it, ignores the basic fact that the system cannot exist, even for a second, without God. God is much more directly involved in the continued existence of the universe than that picture suggests. The laws of nature are themselves the regular and principled ways in which God acts to project a new future out of every present. But is God's action limited to implementing such general principles?

God's action in a universe without persons may be confined to ordering or selecting such general principles. Chaos theory shows how, if they are carefully chosen, such principles can generate infinite complexities of order and patterns of beauty. Perhaps God realises new particular values by generating them in accordance with law-like principles. On the other hand, there could be cases of unique patterning, not repeated throughout the universe, in which elements are ordered in unique patterns of beauty, within the general structure of law-likeness and randomness that the cosmos requires.

Because the cosmos is intended to be self-shaped, it must allow scope for the emergence of beings capable of discipline, effort and finite creativity and, accordingly, for their contraries: egoism, non-mindfulness and the conflict caused by greed, hatred and delusion. Such self-shaping will only arise when consciousness has evolved, when beings have developed a sense of self that they are able to discipline or not. Such self-discipline is intended to lead to fully personal relation to God, to the free co-operation of creatures with God, in responsive interaction. That means that the physical structures of the cosmos must allow for a genuine mediation of the divine presence, for the generation of material sequences that encode various expressions of the divine nature, to be 'read' by finite minds.

The physical structure must accordingly allow for improvement or frustration, depending on the actions of finite agents, and it must have an incompleteness or openness that will allow for creative and responsive divine ordering, within the limits of intelligibility and chance. Moreover, from the first moment of this temporal universe, God does need to ensure that finite agents do come into existence, and so one may think of God as exerting a continuing constraint or influence on the way physical processes go. The system must be such as to allow many possibilities of divine influence, as long as they leave the general structure intact. In general, one might suppose that God might often or even continuously act in unique and local ways to realise particular forms of beauty. But God will generally ensure that the general laws of nature continue to exist, and that the limits of probability are not broken.

From a Christian point of view, the whole universe is creatively shaped by the divine Spirit, on archetypes present in the eternal Wisdom or Logos of God. It is no blind process, without any goal. It is intended from the first to attain a goal, and it is directed with supreme wisdom to that goal. The Spirit will act through the open, law-like, emergent and holistic physical system of the universe in ways that direct it as efficiently as possible towards its intended goal. Such acts will be physically undetectable, as they are not the result of physical forces, and they are compatible with the continuance of all known laws of nature.

In consequence, it will always be possible to suppose that things happen solely according to the laws themselves, plus perhaps some indeterministic factors. One will never be able to observe God at work, since the Spirit is unobservable. What we can do, however, is to calculate the probability of the laws producing the outcomes they do produce on their own, and compare it with the probability of such outcomes on the hypothesis of divine influence, however unobservable.

Such a calculation is not easy, since we do not know the precise constraints on divine action, or the precise purposes of God at any time. Christians believe, however, that God wills the existence of rational beings, so we can say that the probability of their evolving, by however circuitous and complex a route, is as high as it could be, in fact it is certain. We also know that there are vast numbers of possible outcomes to any evolutionary process, so that the prior probability of any one of them ensuing is quite low. Since rational beings do exist, it looks much more probable that God should have influenced events to produce them, than that they should have evolved by blind chance.

One should not think of God, as Newton admittedly did, as nudging physical processes very occasionally, when it looks as if they are going to break down. Rather, God is an agent who is continually seeking to influence physical processes in appropriate ways, as physical constraints allow. In the organisation of quarks to form relatively stable atoms, in the binding of atoms to form enduring chemical substances, in the positioning of a planet in a solar orbit that permits an atmosphere and large masses of water and carbon elements to interact, in the formation of immensely complex self-replicating molecules, in the finely tuned physical interactions that produce mutational changes in DNA, in the structuring of the environment that favours the selection of more complex organic forms, in the organisation of living cells into co-operative structures of the body, and in the emergence of central nervous systems that make possible consciousness and action – in all these things God is a causal influence patterning physical events in specific ways, giving them a tendency to complex and consciousness-oriented organisation.

One might think of the Spirit as a co-operating influence for good, never obliterating the proper autonomy of nature, but conforming it, slowly but surely, to patterns laid down in the eternal Word. Plato basically had it right, in seeing the Forms as the archetypes of all things and seeing the cosmic Designer as fashioning physical realities upon those archetypes (in his dialogue the *Timaeus*). The Christian vision places the Forms securely in the Word of God, identifies the Designer with the Spirit of God, and insists that matter itself is the creation of God, but a creation with its proper autonomy. That autonomy is determined by the nature of its goal, which is that it should be taken up into conscious relationship with God, through the mediation of finite persons, co-workers with God, and become the sacrament of the divine life.

The Spirit of God is not an occasional interferer with nature, or a divine compensation for the failure of physical laws. The Spirit of God is always and everywhere exercising her creative influence, responsive to every change in the physical order, seeking to work towards the realisation of the eternal Word, in his infinite forms, in the temporal world. The Word of God is the all-inclusive form of all that can possibly be. The Spirit of God is the shaper of time into the forms of eternity. Together the Word and the Spirit lead creation towards the completed community in which finite beings can enter into fully conscious and loving relationship to God.

9

God and Evolution

The Process of Evolution

Modern science sees the whole cosmos as subject to processes of evolutionary change. From the moment of the Big Bang to the fantastically complex organisation of the molecules of DNA, new physical structures have continually emerged, in ever more complicated forms. It seems hugely improbable that, in the primeval seas of the planet earth, amino acids should meet and combine to form large molecular structures capable of self-replication. It is even more improbable that long strings of DNA should coil into the nuclei of cells, and that cells should differentiate to produce organisms, colonies of cells which co-operate to form limbs, mouths, digestive systems and bodies. The motive for positing some sort of intelligent design is almost overwhelming.

It is equally improbable, however, that an intelligent and all-powerful God would have directly planned and executed this design in every detail. There are just too many mistakes and dead ends for that. A God who planned every detail would never have made an anacephalic child, a child born with no brain, because of some copying error in the transmission of DNA, some genetic defect. An all-determining God would have allowed no genetic defects.

One must therefore look for a design that is not all-determining, that gives a sort of general blueprint, but leaves many details to be filled in later, in many different ways. The system could be set up so that physical structures inevitably in the long run give rise to organised complexity, without laying down exactly how this is to be accomplished. One model that seems plausible is the model of an unconscious purposiveness in natural processes, by which they move towards a goal that they do not consciously apprehend. Plato thought of material things as partly caused by ideal Forms that could only imperfectly be instantiated in matter. Such an idea seems alien to modern science, which has done its best to discard all notions of teleology and of real natures, replacing them with ideas of unconscious law-like processes. But when those processes result in structures of such organised complexity that it looks almost impossible that they could ever happen at all, it may be time to think again. A theist will normally believe that God does envisage the sort of conscious persons God intends to create. There are ideas of such persons existent in the mind of God. Could those ideas have a causal influence on the physical world? They would not be determining causes, bringing into existence physical realities exactly as God intends. But they could be selective causes, selecting from what becomes available by the natural processes of evolutionary development, pathways that are better suited to realise the eventual goal.

Of course, God sets up the natural processes that generate new possibilities through time. These processes are structured to provide an open, emergent, holistic and stochastic physical system, within which free, autonomous, rational agents can evolve. The evolution of life on earth gives the best example of such a process. It may seem largely random, yet in fact the planetary ecosystem provides a finely balanced environment for the development of conscious life-forms. God's selective causality will be exercised within the constraints of that probabilistic ecosystem. This means that the sorts of selection open to God at any time are not knowable by us, and therefore not predictable or observable by us. One would expect, however, that such divine causality would operate both in the processes that bring self-replicating processes (life) into being and in the evolution of biological forms from simple to extremely complex. A materialist may say that undetectable causal influences should be ignored, especially if they are not strictly necessary to explain what happens. But anyone who believes in God is committed to the belief that there are such influences, if God ever does anything at all. They will be physically undetectable because God is not a physical existent, and so cannot be observed to act, as human beings can. How, then, would one ever know if God had acted?

Unless we knew all the causal factors involved in every change, it would be hard to say whether physical factors alone accounted for such change or not. The fact is that we are not in a position to know all causal factors. No honest scientists would say that they had exhaustively catalogued all causal factors in any situation. If God acts in such a way that no physical laws are broken, such action would be imperceptible. But one might reasonably suspect it was present if things changed in ways that would look very improbable on chance alone, but quite likely given intelligent 'nudging', or selection within the parameters given by the combination of intelligibility, randomness and emergence. The evolution of consciousness on earth does seem to be very improbable, and wholly unpredictable, by purely random mutation. But it would become highly probable, and still remain within the laws of nature, if selection within the possible mutations available at any time was influenced by an intelligent consciousness.

So one can think of God influencing the universe within the limits of physical laws. There is a problem, however. If God actually influences the material process to cause it to produce rational beings, it must look as though that is what the process actually tends to do. Then we might say that the process on its own tends to produce such beings, and we do not need to introduce God at all. How can one tell whether there is a 'hidden selector' at work in the evolutionary process, guiding it towards some pre-envisaged goal?

The hypothesis of God is not itself a scientific hypothesis. It is not needed, even in evolutionary biology. Indeed, it would not help an evolutionary biologist at all to appeal to God, since one would then in effect be giving up and saying that there was no natural explanation for physical processes. There is good reason to keep God out of biology. We do not want a lazy appeal to the divine will to hold up empirical research. All the evolutionary biologist needs to do is to appeal to the four great principles of the neo-Darwinian theory of 'modification by descent', and show how they illuminate the development of organic life from primitive prokaryotic cells to human beings.

These four principles are: 1) the principle of replication, according to which the genome or DNA sequence is replicated profusely; 2) the principle of random mutation, according to which the genome will differ in small, random but sometimes phenotypically significant respects, either because of copying 'errors' or chemically caused mutations in the sex cells; 3) the principle of the struggle for scarce resources, according to which organisms compete for existence, and the better adapted survive; and 4) the principle of natural selection, according to which adaptive mutations are conserved through preferential replication, and the process of adaptation continues over millions of generations, 'guided' by environmental pressures.

Armed with these principles, one can give an illuminating account of how self-replicating molecules could arise and, over a time span of about one and a half thousand million years, develop into a tree of life with many branches, each growing by random mutation, most of them ending in extinction, but some continuing to be selected by the environment, and producing all the many variously adapted species that exist on earth today. There, one might say, the task of the evolutionary biologist ends. One sees how the many species of organic life have evolved by random mutation and natural selection over a huge number of generations.

The question of God only arises once this story has been told. Then one may wish to know if this is a process that God could have instituted, or in which God might be actively involved. Or could the process exist equally well without God, so that the existence or non-existence of God is irrelevant to the process? Or is the process perhaps incompatible with at least the sort of God, a God of universal love, in whom Christians believe? These are the fundamental questions with which the theory of evolution by natural selection confronts Christian believers.

Is the Evolutionary Process Wasteful or Random?

Some biologists hold to an incompatibility view, and that is because of the nature of the process as they see it. The process is, they say, both wasteful and random. Vast numbers of species become extinct, and it seems to be sheer chance that any species survive at all. The process does not seem to have any goal or purpose. It involves so much death and suffering – one species eating another, nature being largely dangerous and hostile to life, that one cannot think of there being any purpose or of a loving creator. We are, in this view, immensely improbable accidents in a pointless universe, with many ill-designed features, but well enough adapted to the environment – though only temporarily – to have survived so far.

These judgements are not entailed by the biological theory of evolution, but are evaluations of the process of evolution, and it is not difficult to see that a very different set of evaluations is available. 'Waste' is a good case in point. I have not had any children for some years, so as far as natural selection goes I am a useless by-product of the process. I am so much waste. But I would not consider my life wasted. Nor would I consider that the only point in my life was the production of my children, wonderful though they are. So why should we consider all the organisms that do not produce offspring, all the species that become extinct, a waste? The dinosaurs were not a waste of time, just because they, or their descendants, did not live for ever. They have a value that lies in their sheer existence, whether or not they left descendants, even if they all died out. They probably had many sorts of experience that were worthwhile, so their lives were of value to themselves. And their diverse and fascinating forms would be of value to God, who could appreciate their distinctive forms of life. We would only call the evolutionary process wasteful if we assumed that it only had one point, which was for everything to have descendants. But what if the point lies in the complex and varied forms of life that it produces? Then the production of a great many varied forms will not be a waste, but a triumph of the process. The dinosaurs were a success, not a failure. So it is clear that what the evolutionary pessimist calls 'waste' will be seen as a fascinating and unique creative moment of the life process by an evolutionary optimist.

What about the randomness, the sheer accidentalness, of the process? That again is a consequence of evolutionary pessimism. For a start, it is not clear what is meant by 'randomness'. In the human genome there are estimated to be 3 times 10 to the power of 9 nucleotides. The number of possible combinations of these nucleotides is astronomical. If one ran through them as quickly as one could in random order, the whole history of the universe would not give enough time to complete the sequence. The chance of hitting on the genome sequence for a viable organism by such random shuffling is vanishingly small. It should be noted, too, that even such a shuffling would not be truly random. It is limited to a specific number of nucleotide bases – four to be exact. And one would need a rule to preclude getting into an infinitely repeating shuffling loop, to ensure that all the possibilities were reached sooner or later. This would in fact be a closely ordered shuffle through a finite selection of physical possibilities – not at all random, after all.

Even then, there are further weightings of the possibilities, dependent upon the stability of chemical bonds between the nucleotides and the ordering of these bonds into codons, or triplets, which code for specific proteins. In fact, when one looks into the physical basis of DNA, one may begin to wonder if there is anything random about the process at all. It could be determined by underlying physical laws, too complex for us to discover. Most physicists are agreed, however, that there are no underlying determining laws, 'hidden variables', waiting to be discovered. Yet it is important to note that the processes are not just random. They are highly ordered, even if the laws that order them are stochastic in nature, rather than determining.

The most widely held scientific view is that there is some randomness in the mutational process, but that it is rather finely tuned randomness. Once a genome sequence gets established, mutations do not happen haphazardly, or by absolutely huge, destabilising jumps. There are paths of genetic change, and near neighbours of actual genomes are much more likely to occur than sequences that are far removed. Most biologists agree that genetic mutation is not 'directed'. That is, it is not weighted so that more favourable changes occur than unfavourable. Nevertheless, it is incontestable that genome mutations throw up favourable adaptations often enough for them to become well established in the gene pool. Since one cannot know in advance which changes will be adaptive, one could see these so-called 'random' mutations as giving plasticity, indeterminacy and thus flexibility for the production of new types of organism. In an unknown and changing environment, mutation within closely defined limits of rigidity and fluidity will provide a high potential for adaptation, for conservation of adapted forms, and for the emergence of ever-new adaptations.

Furthermore, as one looks at the sort of phenotypic changes that genetic mutations produce, it seems that there is a tendency to develop the same sorts of adaptation quite independently. The eye, for instance, has evolved independently a number of times. In general, mutational changes in the genes have brought about the development of multi-celled bodies, limbs, senses and nervous systems where there is no particular reason for thinking that such developments would even be possible, given just the basic laws of physics and chemistry.

This can be seen as a process that does contain randomising elements – the small, controlled genetic mutations that produce phenotypic changes, some of which are selected by the environment. But these randomising elements are themselves quite finely tuned to produce such changes, and they exist within sets of physical processes – environmental factors, laws of physics and chemistry – which are not at all random, but physically well controlled.

When one considers that mutations are themselves partly the result of environmental influences – in the form of cosmic rays or radiation, for example – and that they result in phenotypes, some of which give rise to sequences of accumulative complexity in their environment, it becomes plausible to see mutation as just one element of larger-scale physical processes in a wider ecosystem. Then, 'randomness' begins to look more like a device for generating flexible and creative possibilities from which the changing ecosystem can select the most promising candidates than like a long series of little blind leaps into the dark.

Pessimism and Optimism about Evolution

Is it an accident that human beings developed on earth? Evolutionary pessimists hold that it is an accident, in the sense that there are many other possible paths that mutation and selection could have taken, so far as we can tell. Most of them would have led to the extinction of all life, to evolutionary dead ends. Perhaps the dinosaurs were an evolutionary dead end, very efficient predators that prevented the development of any self-conscious organisms by eating them before they got going. So it is very improbable that self-conscious beings would have ever evolved, and the fact they have evolved on earth is largely due to the accident that a meteor wiped out the dinosaurs sixty-five million years ago and allowed mammals to develop along a new evolutionary path. We got here by accident, and if we ran through the process again, it is most unlikely that we would come into existence a second time.

Again, however, one can interpret the same process in a very different, much more optimistic way. Of course the development of self-conscious life-forms is very improbable, given all the other paths evolution might have taken. In fact, one can raise this improbability immensely by considering all the 'accidents' or 'coincidences' that needed to occur to develop human beings. The extremely complex structure of DNA molecules might never have happened, and might never have coded for the building of proteins that construct well-integrated phenotypes (bodies). Most sequences of nucleotides would, after all, produce gibberish, which would not build anything, and the chances of hitting on the comparatively tiny number of sequences that do build coherent bodies are vanishingly small. It is just amazing that any structures ever came to self-replicate at all.

Then it is highly improbable that the mutations that occur in the genomes would give rise to graduated phenotypical changes, exhibiting the plasticity that renders them adaptive for changing environments. One would expect most mutations to be harmful, and it is an incredible piece of luck that small changes in the sequencing of nucleotides should cause such things as the development of eyes and limbs, in ways that are not too great to destroy the organism and not too small to have no adaptive effects. Mutations might not be 'directed', but the sorts of phenotypic variations they produce are just what is needed to make possible the pursuit of graduated and continuous paths through genetic space. There seems to be no a priori reason why this should be so.

The evolutionary process further requires that the ecosystem should be friendly to some organisms – that there should be a planet at just the right distance from a star to provide the possibility of an atmosphere, an ozone layer to screen harmful radiations but open enough to allow radiations that produce some genetic mutations, and a gravitational field suited to allow complex organic bodies to form. Organisms are 'naturally selected'; that is, the environment selects those that are best adapted. In fact, the environment seems to foster the development of moving, perceiving organisms wherever mutation produces them, and to impede the development of purely parasitic destroyers. The dinosaurs, for example, were selected by the ecosystem as good perceivers and agents, but for various reasons they proved to be an evolutionary dead end. The mutational changes available to them were perhaps not capable of providing further phenotypic changes, and the ecosystem (which of course includes the meteors and planets, and even the furthest galaxies in its scope) rendered them extinct. Then the ecosystem selected a new set of mutating organisms, which ended with *Homo sapiens*. The fact that the ecosystem as a whole has been friendly to the development of *Homo sapiens* is thus fantastically improbable, given all the other possibilities some of which, apparently, almost occurred.

Finally, the evolutionary process shows a cumulative adaptation, an increase in complexity and consciousness, from unicellular organisms, through plants, the lower animals, to self-conscious and self-directing agents. It is not human arrogance, it is sheer fact, to say that the human brain is the most complex integrated structure on this planet, and shows a huge increase in coherent structure over all other organic forms. Each step on the pathway to the development of the neo-cortex is itself improbable. So the whole pathway is cumulatively improbable, to a huge degree. Long sequences of small 'improvements' in organisation do not increase the probability of the sequence, but render its existence more and more unlikely, the longer it exists.

In all these ways, one can agree that the evolution of beings with a neo-cortex is vastly improbable. In fact, it is just too improbable to be an accident. The evolutionary pessimist says that vast improbability shows human life to be an accident in a purposeless process. But a more plausible reaction would be to say that vast improbability amounts to impossibility, unless the whole process has in fact been fixed. If the very unlikely happens, a good explanation is that somebody has made it happen, has ensured that it will happen, against all the odds. As far as evolutionary biology is concerned, one can be happy just to say that it is a very improbable process, but it has happened, and one has to deal with that. But if one is asking the non-biological question, whether a very improbable process is compatible with intelligent design, the answer is that if the process is elegantly structured to a good end, then the more improbable the process, the more likely it is to be the product of intelligent design. The argument that the evolutionary process is incompatible with design misses the mark completely.

Theistic Evolution

The last move the evolutionary pessimist, or materialist, can make is to say that if the process was designed, it would be more efficient than it is, or would issue in a state that was better (had less suffering, for example) than it is. Inefficiency of structure or imperfection of the goal counts against intelligent design or purpose. Once again, what is at stake here is not a matter of biological theory or observation, but an evaluation of the sort of efficiency and perfection the world exhibits. It is not in dispute that evolution shows a cumulative adaptation towards the existence of conscious self-directing organisms in a highly fine-tuned ecosystem. It is not in dispute that this is highly improbable, that it could have occurred by chance, but that it is more likely to have occurred by intelligent design. What is in dispute is whether a postulated benevolent designer could reasonably have produced such a process, or whether that would not itself need just as much explanation as the system. I have dealt with the latter point. The remaining question is: is the system more inefficient or imperfect than it need have been? Is it choosable by a benevolent God?

It is entirely plausible to hold that the system is highly efficient, if the aim is to produce autonomous, self-shaping material creatures, within an open and emergent system. Physicists are more likely than biologists to note how superbly the basic laws and constants of nature are tuned to be just what they need to be to produce carbon-based life-forms. When biologists talk of inefficiency, they are thinking of the evolutionary blind alleys and harmful mutations that follow from the indeterminacy of the evolutionary process. But if the whole point is to introduce that finely tuned indeterminacy that is inseparable from the development of creaturely freedom, then the dead ends come to seem a positive part of the richness of the life process, and the harmful mutations are unpreventable consequences of a randomness which is necessary for the emergence of creativity, freedom and the realisation of fully personal relationships within the universe. The system is highly efficient for its purpose, which is not the immediate production of fully fledged perfect beings, but the emergence from material structures of self-aware and self-directing communities of agents that can themselves transform those structures into vehicles of conscious mind.

The idea of perfection is very hard to pin down. At first sight it seems more perfect to be supremely happy, knowing and powerful than to be capable of suffering, ignorance and weakness. But there is a distinctive sort of value in winning happiness through disciplined effort, through the exercise of distinctive capacities. There is a value in learning for oneself the secrets of the natural world. There is value in having to co-operate with others, and thus in being less than maximally powerful. For human beings, the pursuit of truth, through investigation, reflection and the overcoming of partiality and prejudice is a great good. The pursuit of beauty, through the discipline of attention and the honing of creative skill through a personal expression of one's own potentialities, is a great good. And the pursuit of goodness, through learning to co-operate in trust, loyalty and compassion, is a great good. But all these goods entail the possibility, and to some extent the existence, of ignorance, failure and frustration, and conflict.

The theist will say that the universe has not yet attained its goal, which is one of fellowship, understanding and the love of beauty. But the evolutionary movement towards this goal, though to some extent frustrated by human evil, has been sufficiently successful to justify the faith that God has designed evolution to realise forms of goodness which otherwise could not exist, and which are supremely worthwhile. The evolutionary pessimist will say that human life is a pointless accident in an indifferent universe, and that one just has to bear it as well as one can. The evolutionary optimist – the believer in God – will say that human life is a key part in the glorious realisation of rich and complex patterns of beauty and intelligibility. It has a purpose, and is part of a cosmic purpose, to fashion the universe into a temporal image of the divine life. If that purpose has been partly derailed by past human evil, God will ensure that it will be fully realised. In the meanwhile, human existence is such as to allow every human being to begin a personal journey towards infinite love. Of course, that is a religious faith and cannot be given by the science of biology. But such a faith is compatible with the facts of evolution as biologists understand them. This is a system that could well have been designed by a benevolent God, for forms of goodness uniquely realisable within it.

So the process of evolution could have been designed by a benevolent God. But could it equally well have occurred by chance, so that the hypothesis of God is superfluous? Or must God have played an active part in the process, directing it to take the improbable paths it has taken, on the way to the development of self-conscious life?

It must be accepted that it could all have happened by chance. Further, the biologist would do well never to appeal to God as a designing power, since such an appeal would take one outside the realm of formulable scientific laws. However, that does not mean God is a superfluous hypothesis. As I have noted, the existence of God renders the process, as it has happened, much more probable than it would otherwise be. So it is a good metaphysical hypothesis, if not a good scientific one. The atheist will be unmoved by this, insisting that one does not need to appeal to God, and that God is anyway a contentious entity to introduce. But the theist, who already has a belief in God based on other, experiential and personal grounds, will equally reasonably be able to say that the existence of God is not only compatible with evolution, but makes its history much more probable. For the theist, this will be an additional confirmation of the existence of God, and there is no need to put it forward as a conclusive proof. It will, however, help to show that belief in God is entirely reasonable, consistent with the best scientific knowledge, and plausible as an underlying, non-scientific (not experimentally testable) explanation of why the universe is as it is.

It is impossible either to prove or disprove the activity of God as a selective cause in the process of evolution. We do not know the deepest causal structures of nature. But the theist will probably want to deny that God keeps interfering in an otherwise stable structure to stop things going wrong. One might rather say that God is the ultimate basis of the structure itself, so that the directing of its progress towards consciousness can truly be seen as an internal constraint or set of constraints on the physical system. The theistic view that this series of constraints is consciously and intentionally directed is coherent and plausible. In this sense, the theory of evolution gives valuable new insights into the relation of God to the created universe.

10

The Soul

The Emergence of the Soul

The Christian view is that one of the chief goals of creation and evolution is the emergence of beings that to some extent possess awareness, creative agency, and powers of reactive and responsible relationship, with whom God can enter into personal fellowship. The universe is ordered from its beginning to the actualisation of beings made 'in the image of the creator'. Human beings, in this view, are not accidental by-products of blind cosmic processes. They are parts of the envisaged and predestined goal of the evolutionary process. The existence of consciousness, purpose and moral agency is not some strange and temporary anomaly in a ceaseless recombination of atomic parts. It is that for the sake of which the whole material process has been laboriously and intricately constructed.

The philosophy of materialism, which regards the existence of a spiritual (immaterial) God as impossible, and which attempts to account for everything in terms of the motion of material elements, runs up against its hardest counter-examples at this point. As human beings, we are aware of sensations (touches, smells, colours, sounds), of images (as in dreams), of feelings (of joy, regret, desire and happiness) and of thoughts (beliefs, memories and purposes). These are the contents of our consciousness, and they are known to us more immediately than anything else. They are the starting-point of all our knowledge.

It is the relation of the contents of consciousness to the physical and chemical structure of the universe that constitutes a major problem for the materialist worldview. We can fairly easily see how quarks can combine to form protons, neutrons and electrons, how subatomic particles can combine to form atoms, which combine to form molecules, cells, organic bodies, nervous systems, and brains. There is no problem, in principle, in seeing the brain as just a particularly complex organisation of quarks. But there is a huge problem in seeing my understanding of a complex argument, an act of human understanding, as just a complex organisation of quarks. The problem is that this seems to be a different type of thing altogether. It belongs to a different logical category. How can one be 'made of' the other? How can bundles of quarks, however complex, understand anything?

When we think about visual perception, we can trace the reflection of light from the surfaces of objects to the retina of the eye. We can follow electrochemical impulses from the eye along the optic nerve to the visual cortex. We can pinpoint the location in the brain of electrical activity which causes us to see an object. It may even be possible to obtain an exact correlation between the occurrence of the firing of specific neurons at a number of sites in the visual cortex and my report of seeing a brown desk in front of me (though present research suggests that there is no such one-to-one correlation of brain-states and visual reports).

If everything is wired up properly, if my brain gets into these electrical states, then I will see a brown desk. If, by some artificial stimulation of the cortex, identical brain-states were caused to exist, I would still have the experience of seeing a brown desk, even in the absence of any external world. Thus it seems a well-established hypothesis that conscious experiences of perception are caused by the occurrence of physical brain-states. The perceiving itself, however, the awareness of the presence of a brown desk, the understanding that such a desk is present, does not seem to be a physical state.

Physical states of the brain can be observed and measured. One can record the electrical activity in the neural networks of the brain and give it location and intensity. But from none of this observational data could one have any idea of what it is like to perceive a brown desk. All one can do at that point is to appeal to introspection, to one's own experience of perception. One might then argue by analogy that beings with sense organs and brains rather like ours are likely to have experiences something like ours.

That is what we do when we try to imagine what sorts of experiences a dog, for instance, might have. We know that a dog has a much greater area of its brain devoted to the sense of smell, we know the sorts of molecular structures that give rise to smells, and we have a rather rudimentary sense of smell ourselves. Thus armed, we can try to imagine a dog's perception of its environment, though obviously we can never know with how much success. In particular, since we have a hugely more developed neo-cortex, it is hard to tell how much that affects the character of human experience. Descartes' supposition that animals are not really conscious is not wholly absurd, since it could be that full consciousness depends on higher brain activities, which are not generally present in at least most animals.

On the whole, however, neurophysiologists tend to stress the similarities between animal and human brains, and the continuous line of development from the tiny ganglia of insects to the neo-cortex of humans. We know that organisms are capable of responding to their environment, and acting within it to obtain food or escape capture, without any form of consciousness. We know that because we can build robots programmed to do precisely those sorts of things, and we are sure they have no consciousness. We also know that we are conscious of things around us, and can act purposively and responsibly. Accordingly, we know that at some point in the development of the brain some primitive form of awareness and intentional agency must arise as a new emergent property. We are just not sure of where exactly this is.

Does this mean that at some stage in the development of the brain a non-physical entity comes into existence, which is dimly aware of its environment, a little proto-soul? Perhaps one will not wish to speak of an 'entity', but it is not absurd to suppose that non-physical – experiential – properties come into existence when a physical structure reaches a critical point of integrated complexity. The experiences of lower animals could be sequences of qualia, or sense-images, dimly apprehended as desirable or undesirable, and evoking muscular responses almost automatically. Such experiences, which one may imagine to be rather like dreams, will be non-physical properties generated by the primitive brain, and remaining causally dependent upon that brain. Primitive consciousness will be a property of a physical brain, or a series of such epiphenomenal – dependent and transient – properties. This idea roughly corresponds to what Aristotle called the 'animal soul', the power of sensation and movement that characterises animal life. Such non-physical properties have no substantial or independent existence. They are sequences of images, bound together by their common causal relation to one particular brain.

At the other end of the continuum of developing consciousness, the developed 'rational soul' of *Homo sapiens* is a structuring and organising principle that interprets such images as appearances of an enduring world of objects, whose nature it can come to understand in terms of abstract concepts. It will order its activity to obtain envisaged future goals within that world. It will have a sense of its own continuing existence, with a responsibility for shaping its own future, and a concern for the quality of its experiences and for the experiences of others. It will regard itself as one continuously existing responsible agent among others in a moral community.

The rational soul supervenes upon the image-sequences of primal consciousness, and is able to achieve a sense of understanding, purpose and responsibility. But what is its relation to the brain? Primal image-sequences are causally generated by brain-states. It may seem odd that physical states can cause non-physical states, but causal relations just have to be accepted for what they are. A theist will say that God has set up the system so that non-physical states will be generated from physical processes, at a foreordained level of complex organisation.

Is the rational soul generated in a similar way? At some point in animal evolution, the dream-like images of primitive consciousness begin to come under the control of a subject with the power of abstract understanding and responsible action. It is at that point, whenever exactly it is, that the rational soul begins to exist. The rational soul emerges from a complex material basis in the course of evolution. But, as a subject of comprehended experience and responsible action, it transcends its material basis, and realises a radically new form of existence.

The Embodied Soul

The rational soul is causally dependent upon the existence of a highly developed neo-cortex. It comes into existence with the genesis of such a normally functioning brain. But it cannot be simply identified with the physical structure of the brain. Nor is it a quite separate entity, as though, having been generated, it could then exist in isolation from the brain. As a subject of understanding, it requires data to understand. As a subject of action, it requires a physical environment in which to act.

One way of representing the situation is to think of the brain as a library in which information is stored in coded form, rather as compact discs can store information in strings of binary code. The 100 billion neurons of the brain store electronically coded information, rather as hard discs in computers store information in coded form. Such codes, however, must be 'read' by a suitable interpreter. Without the reader, the codes are meaningless – just as a compact disc means nothing without a CD player. Without the coded information, there is no content to be read. Human beings need a properly functioning brain to provide information content that can be referred to repeatedly, if necessary. Malfunctions in the brain – loss of memory cells, for example – will entail a loss of information content. But all that information needs a reader/interpreter who can be consciously aware of it. This reader is the rational soul.

This makes the soul sound as if it is a passive reader of whatever information the brain provides. It cannot help observing whatever is written in the relevantly open books in the library of the brain. But the soul in fact has a primarily active role. It can cause changes in neuronal activity in the brain, which will in turn cause bodily movements and bring about changes in the environment.

When one considers the 'reading' function of the soul, it turns out to be not so passive after all. At relatively simple levels, I may just record the occurrence of brown patches in my visual field. But, in fact, perception is a complex activity, in which memories, thoughts, intentions and various forms of attention or inattention are concerned. This is not just any brown desk. It is a familiar, friendly, well-known brown desk with a history and a defined role in the world of my daily activities. It calls up, not just coloured patches, but remembrances of tasks achieved, reminders of tasks unfinished, feelings of struggle and success, of affection and despair.

So the electrical activity in my brain is not a one-way causal flow. Input from the senses and from other parts of the brain merges with a repatterning and reconnecting activity in a continuous receiving and modifying of information, all intermingled in a seamless reception of and response to new stimuli. The soul is the dynamic focal point of all this reception and response, embedded in the pulsating web of electrical energies that comprise the activity of a living brain. It is the co-ordinator and executive director, not wholly bound by the physical processes of the brain, yet always working within the brain's physical constraints and in response to the physically encoded information the brain provides.

If this is a plausible account of consciousness, it follows that the soul is not itself a material or physical entity. Some philosophers have called it a 'spiritual substance', different in kind from physical substances. This may be rather misleading, however, if one thinks of a substance as a continuing and relatively self-sustaining entity. If the soul is primarily the reader/interface with the brain, it needs the brain in order to act as it is meant to act. Without a brain, there would be no source of information and no instrument to act on. So the soul does not have a truly independent existence. It is parasitic on the brain, in a very real sense.

Why, then, should one not call the soul an emergent by-product of a properly functioning brain? Might the soul not be just the characteristic way that a complex physical entity like a brain acts? There is nothing wrong with calling the soul emergent from the brain, if that means that whenever a brain functions properly, then a soul will be in existence. But the soul is not just the way a brain acts, as though it were just a peculiarly complex pattern of physical activity. The soul understands, responds and acts. It reflects, contemplates and decides. The soul is an agent, and it is the same agent, continuing through a period of time. It is a continuing agent, essentially related to some instrument of information and interaction with an environment – essentially related, that is, to some brain and body.

Human beings are essentially embodied souls, continuing agents that are embedded in particular brains and bodies. They have consciousness, whose content is provided by a particular brain. They can devise purposes, which they enact through using the potentialities of a particular brain. They have moral agency, relating to other souls by continuing responsive interaction with their embodied forms. The relation of soul and body is very intimate, so intimate that each soul is appropriately fitted for just one body, and each body has the capacity to be the express image of a particular soul, in a community of embodied beings in continual interaction.

Human beings, then, are more than just bundles of physical particles bound together in complex ways. They are subjects of understanding and responsible action, and that subjectivity always transcends any purely physical understanding and eludes experimental and empirical observation. It is the core of human personhood, which gives humans their peculiar dignity and distinctiveness. Yet humans are properly embodied beings, and each unique subject reflects and acts in the world through the brain and body that is its material expression. It would not be better for humans to exist in disembodied form, to escape altogether from physical embodiment. It is precisely the physical world and its beauty and intelligibility that they are fitted to understand. Within that material world, they are called to realise new forms of value and relationship, by responsible action.

The Christian view of the human soul is thus quite different from the sort of Platonic dualism which holds that the soul is trapped in the body, and would exist in a purer form without the body. The aim of Christian life is not freedom from the body, or liberation into some purely spiritual realm in which individuality and time cease to be of importance. Christians believe that God has created the physical world, so that it is inherently capable of goodness. God took incarnate form in the physical world, so that it is inherently capable of manifesting spiritual reality in an adequate way. The Spirit of God works throughout all creation to bring all things into reconciliation in Christ, so the destiny of the physical world is to be transformed into an appropriate expression of the glory of the creator.

The human body is part of the physical cosmos which, through long ages of evolution, has become fitted for embodying finite consciousness and will. Through the human brain, the universe is known, not as God knows it, omnisciently and inclusively, but from the restricted viewpoint of one spatio-temporal location. It begins to be shaped, not by the unlimited power of the creator, but by the limited and laborious exertions of a physically embodied agency. It begins to manifest, not the perfectly executed plan of a supreme intelligence, but a hesitant and ambiguous advance towards global co-operation and understanding, in a world torn by selfish and short-sighted conflicts.

The tragedy of the human situation – and it is a real tragedy, which not even the divine redemption of the world can eliminate – is that the human advance towards truth, beauty and goodness, which should have been undertaken in partnership with the divine intentions for the cosmos, has been rendered virtually impossible of attainment. The body is not, as in Platonism, the tomb of the soul. But all too often the body so cramps and limits the proper activity of the soul that the soul's true nature and potentialities can hardly be recognised.

For the millions of human beings who live under conditions of starvation, brutal oppression and torture, it does not ring true to say that their bodies are the proper embodiments of their souls, the appropriate expressions of their spiritual natures. It may well ring true, however, to say that is what their bodies *ought* to be, that it is a denial of their true natures to subject those bodies to starvation and torture, that it is their very humanity that is being crushed by the physical conditions of their existence.

It is not enough to say that only their bodies are being crushed, whereas their souls are free. To crush the body is to assault the soul, and social and political systems that diminish the bodies and the physical welfare of human beings are attacks also upon their souls. Thus, the Christian view is that souls are properly embodied, and God intended such embodiment to express the many unique forms of understanding and creative action that together comprise a community of persons. That divine purpose has, however, been frustrated by evil, and bodies often limit and impede the expression of the soul's true nature. The basic Christian hope, therefore, is not for disembodied immortality, but for a form of embodiment that will release the true nature of the soul. It is not for the immortality of the soul, but for the resurrection of the body, a fuller material expression of the potentialities of the embodied soul.

11

The Fall and Salvation of Humanity

The Soul in Bondage and Release

Christians believe that when God became incarnate, God did not simply assume a perfectly formed, beautiful and well-fed body, discoursing with the philosophers in shady Athenian groves. Jesus of Nazareth, in whom God is said to have been incarnate, lived a mendicant life, without home or family of his own, mostly in rural villages far from centres of culture and influence, a member of a people under military occupation, and he was tortured and executed as a political subversive. The incarnate God of Christianity identifies with the poor and oppressed, those who are crushed but not defeated by circumstance. The central Christian image of the crucified Jesus portrays the human body as an instrument of pain and humiliation, not as the free expression of a creative soul.

For Christians, however, the cross is not the end of the story. If it were, it would never have become an important religious symbol. The cross discloses two important truths about the human condition. The first truth is that the human corruption of God's creation by self-love is an assault even upon God, and can only lead to suffering and death. When, in the course of evolution, souls came to exist with the power of responsible choice, they were faced with the possibility of learning to co-operate with God, so that they could grow in wisdom, moral awareness and understanding. However, their evolutionary heritage also placed before them the possibility of subordinating these dispositions to the passions of lust and aggression that had been instrumental in making them the dominant species on the planet.

It would be untrue to say that morality is nothing but a set of genetically imprinted behaviour patterns that have been selected because they favoured the survival of certain genes over others. This is a classic example of the aptly named 'genetic fallacy', the mistake of thinking that to show how something developed is to show what it is. Of course morality uses as its basic data dispositions and behaviour patterns that have been selected through evolution. Appeal to natural selection gives an illuminating (but hardly exhaustive) explanation of how some of these dispositions have come to exist. It explains the mixture of lust, aggression, altruism and submission that characterises human nature. The first human beings had a responsible choice between their lustful, aggressive dispositions and the more altruistic, co-operative dispositions that would have led them to grow in the knowledge and love of God.

In addition to such purely moral considerations, there might in the first human lives have been discernments of the will and purpose of God, giving to morality an ultimate authority that overrides self-interest and a faith in the possibility of a community of loving trust in God. From a religious viewpoint, the deepest purpose of human existence is the free development of a relationship of joyful obedience to the will of God, within a community of justice, peace and love. It is that purpose which was rejected when the fateful choice was made of a path of autonomy, of rational self-will, which placed the descendants of those first humans in bondage to self and its consequent conflict and suffering.

The wages of sin, the path of egoistic self-will, is death – not only the destruction of the community of love that was God's intention for this world, but even the death of the incarnate God, who came to his own creatures and was rejected. The first thing the cross reveals is the gravity of human self-will, its fundamental character as deicide, the rejection of God.

The second thing the cross reveals is, however, by far the most important. It is what makes the cross, surprisingly, 'good news' and not simply a message of universal condemnation and death. For on the cross Jesus willingly gives his life 'as a ransom for many' (Mark 10: 45). That is, he gives up his own life so that his followers can be freed from the bondage of selfish desire, and united inwardly to God. In surrendering his life, in this ultimate act of voluntary self-sacrifice, Jesus makes it possible for God to act through his life in a decisive manner, in a way that both reveals the innermost nature of God and begins to effect the reconciliation of humanity to God in a new form.

The innermost nature of God, which is revealed in the cross, is the nature of kenotic, self-giving, unitive love. The new way of reconciliation to God is a way in which humans can share in that divine love by receiving the Holy Spirit into their lives as a powerful creative and transforming force. The good news of the Christian gospel is that humans who are in bondage to selfish desire and death can be liberated into sharing a life of divine love, an eternal life that death cannot touch, and which no power in all of creation can destroy.

The cross is not only the voluntary self-sacrifice of Jesus for the sake of establishing the 'community of the new covenant'. It is an act of God that expresses in a definitive manner what God is seeking to do everywhere, in ways qualified and sometimes obscured or impeded by the many forms of human response, in all their diverse cultural and historical contexts. A feeble analogy might be a composer who is trying to express a particular musical idea, but needs to wait until she can find an orchestra that can play in an appropriate way, where the players are attuned to the music and possess instruments and techniques that have been developed to a certain standard. Perhaps different groups of musicians will be able to express many different and valuable ideas, though not quite the one the composer is above all seeking to express. Perhaps, to push the analogy further, everyone who tries can be assured that one day they will discover what the composer has in mind, and they will then bring to their performance of it the other things they have discovered for themselves. In the light of that fact, one might think that even the orchestra that does express the key idea will have much to learn from all the other musicians, so that the final joint performance will be even richer and more beautiful than was possible beforehand.

This analogy should not be pushed too far, but it seeks to illustrate how Christians would see in the life and sacrificial death of Jesus a point in human history at which God's 'key idea' could be definitively expressed. There the nature of God might be seen in a definitively clear form, and the way in which God will bring all things to final fulfilment might be given paradigmatic expression. This life can be seen as the paradigm image of what God universally seeks to do in every human culture. The cross shows that God's decision to create is also a decision to take on the risk and actuality of suffering. Some such suffering is necessary to the created universe. It arises from God, not by intention but by necessity. God cannot unilaterally eliminate it. Much suffering is not necessary, but God cannot prevent it from arising as a result of the selfish choices of autonomous creatures. But God can – and does – share both sorts of suffering by experiencing them in their full reality. God can seek ways of eliminating suffering, by co-operating with creatures. The Christian promise, given by Jesus' resurrection, is that God can ensure the complete elimination of suffering and evil in future, when the conditions of the created order can be transformed by the free co-operation of rational creatures, and the cosmic goal achieved.

God is not at all indifferent to the suffering and apparent waste and conflict in the universe. The passion of Jesus expresses the way in which God shares in the suffering of the world. The resurrection of Jesus expresses the certainty of divine victory over evil. And the gift of the risen life of the Spirit expresses the co-operation of God, working through human beings, gradually to transfigure the material order into a manifest sacrament of the divine life.

For Christians, the life of Jesus is a historical expression of the eternal nature and universal action of God. At this point in human history, God is disclosed in Trinitarian form, the Son enduring the cross, the Father raising him to glory, and the Spirit carrying his risen life to all who are disposed to accept it in faith. In these events God discloses the divine nature as not only impersonally concerned with the elegance and beauty of the created order, but as also personally concerned to deliver every sentient life from the bondage of self and suffering into a new resurrection order, the community of the divine Spirit, the kingdom of God.

The Restoration of the Divine Purpose

God does not just begin to act in the life and death of Jesus. God did not just decide to save humanity two thousand years ago, and does not confine that saving activity to those relatively few people who are fortunate enough to hear about Jesus. The Christian view has to be that God has been active from the very beginning of this space–time until now, throughout the vast spaces of the universe. Nevertheless, the ways in which God acts will be generally restricted in a number of ways by the nature of the universe that God has decided to create. First, they will in general continue to respect the limits set by the regularities of the laws of nature. Second, human freedom will in general be respected by God, who gives it. God will not often act in ways that prevent human freedom, even when it wills evil, from being exercised. Third, if God wishes humans to be 'fellow-workers' for good, then God must leave some room for humans to manoeuvre. God cannot always be doing the best possible thing in the circumstances, for that would leave humans nothing to do by their contributions but make things worse. If God leaves many outcomes to human responsibility, then God's influence must often be frustrated by the refusal of humans to co-operate, or by their obtuseness in understanding the divine will. Fourth, in an estranged world God's acts may not be primarily aimed at creating a society of justice and peace, but at making the disastrous consequences of evil clear, and at offering a way of liberation from the bondage of self that will train corrupted selves in virtue and re-orient them for relationship with the divine. They may be acts of judgment and discipline as much as acts of blessing and obvious benevolence.

With these qualifications, it would be natural to expect that a creator God would act in particular ways which would aid the implementation of the divine purpose that creatures should know and love God, and which would establish an interactive relationship between creatures and God. Throughout the history of the world, and not least in the world's religions, God has acted in many ways to realise these purposes. But if the nature of God is to be truly known, and the character of the divine action which leads to the human goal is to be clearly identified and appropriately responded to, one might expect that there would be at least one social/historical context that was favourable to such a definitive revelation. Each religious tradition naturally enough thinks itself to be most nearly, though perhaps not exclusively, representative of an adequate expression of the ultimate reality, the final human goal, and the way to it. The Christian faith is founded on the claim that God has acted in such a decisive and transformative way in the life of Jesus of Nazareth. There, Christians claim to discern that the nature of God is kenotic (self-emptying) love (*kenosis*), that the human goal is a renewed creation in which individual lives find fulfilment and true community in the presence of God (*enosis*), and that the way all creatures will eventually take to that goal is selfless participation in the divine love (*theosis*).

It may seem a large jump from the purpose of the whole cosmos to the life of a young Jewish man in a remote province of the Roman Empire. But it is not absurd to see the purpose of the whole cosmos as the emergence of a community, or a union of communities of sentient agents, in which values are created, appreciated and shared, in full knowledge of and co-operation with the cosmic creator. It is not absurd to see human life on this planet as one place – we do not know whether or not it is the only place – where this purpose comes within sight of being realised. It is unfortunately pretty clear that the cosmic purpose has been corrupted on this planet by the acts of our human ancestors. We can see that if God is going to put the cosmic plan back on course, so far as planet earth is concerned, something must be done to liberate human lives from the corruption in which they are entangled, and re-establish union with the divine plan.

Somewhere and sometime in human history, there must be a divine act, or series of divine acts, that accomplishes, or at least begins to accomplish, that liberation from selfish desire which will make possible the existence of a community of divine love. So it is natural to hope for and look for places in human history where divine acts of liberation and reconciliation occur. We might expect that a God of universal love would be working in some way at every time and place, in every religion and culture, towards the fulfilment of the divine purpose. Given the general constraints on divine action, however, not all times and places will be equally favourable for revelatory and redemptive acts of God. Some cultures will develop in ways that may make it hard to develop ideas of a loving creator, perhaps because of their acute perception of suffering. Some individuals will be more receptive to the divine co-operation than others. Some social and personal histories will be more appropriate for conveying a sense of divine love and purpose than others.

What God reveals will depend in many ways on what human histories, temperaments and cultures enable people to discern of God. Perhaps all cultures will be so infected by selfish desire and ignorance that any revelation will be limited and even distorted to some extent. So what God reveals will depend very much on the historical context of revelation, and will always be subject to some degree of human interpretation or misinterpretation.

Christians can hardly claim to be exempt from this ambiguity and restrictedness in understanding divine revelation. I think it would show arrogance of a very high order for anyone to claim that Christians have the one pure and undistorted apprehension of God, whereas everyone else is trapped in error and ambiguity. Nevertheless, Christians do claim to have discerned an important truth about the universal love of God, and experienced something of God's action to establish a community in which the character of the ultimate goal (salvation) is adequately delineated, and in which the attainment of such salvation is assured, in and through Jesus of Nazareth. In the life of Jesus, a wandering preacher and healer in rural Galilee, Christians claim to discern in a paradigm instance the way in which the cosmic purpose is being set back on course on this small blue planet on the edge of the Milky Way.

12

Breaking out of Literalism

The Uniqueness of Jesus

What God wills for the universe, and for the earth in particular, is the existence of a community of persons who will bring the potentialities of the material world to new and fruitful actualisations, who will be guardians of the earth and fellow-workers in realising the creative purposes of God. Such a community is a community that will rejoice in the loving presence and creative power of God, that will flourish under the rule of God, a community where God is king. It could be called the kingdom of God. One of the best-attested facts about Jesus is that he devoted his life to proclaiming that 'the time is fulfilled, and the kingdom of God is at hand' (Mark 1: 15). One of the best-attested facts about the early Christians is that they believed Jesus to be the king, the Messiah (deliverer and ruler), who was to found this kingdom.

An awkward fact about assessing the place of Jesus in the religious history of humanity is that he is not represented primarily as a prophet or great spiritual teacher. He resists complete assimilation into the set of God-enlightened or God-inspired men and women who have deepened the human sense of God and the divine purpose for creation. What is unique about him is that he is claimed to have a quite specific role in the fulfilment of God's purpose for the earth. He is proclaimed by his disciples to be Lord in the liberated community of the divine love. Such a claim is a stumbling block for any attempt to see Jesus as one religious teacher among others.

It is natural and right to see a God of universal love as working everywhere to reveal something of the divine nature and purpose. This leads one to look at all human cultures for signs of such revelation. It is not hard to find such signs. In the prophets of Israel one can discern a growing insight into the justice and mercy of God. In the Qur'an there is an insistence upon the absolute uniqueness and sovereignty of God. In the Upanishads one is directed to look to 'the cave within the heart' to find the presence of that being of wisdom and bliss that is the one truly Real. In the teaching of the Buddha there is outlined a path to that wisdom, compassion and bliss that delivers from evil and suffering. In Confucian teaching, the 'Way of Heaven' is a way of justice, order and the harmony of all things. Even in modern humanist, anti-religious cultures, the concern for human welfare, freedom and equality can be seen by theists as prompted by the Spirit of God in an ambiguous and hatred-obsessed world.

If one was looking for a distinctive insight given by Christianity, it might lie in the emphasis laid upon the forgiving and self-giving love of God, which seeks to save all who are lost, 'not wishing that any should perish, but that all should reach repentance' (2 Peter 3: 9). But there is something else that is distinctive about Christianity. It does not just teach that God is love as a general truth that might be confirmed in general human experience. It seems to be essential to Christian faith to claim that Jesus is the liberator, the Christ, who establishes the kingdom of God. At this point in human history, the followers of Jesus claimed, the foundations of the kingdom were laid, and God acted decisively in and through the person of Jesus to found the kingdom in history.

That claim is an awkward one, because it asserts a sort of uniqueness for Jesus that seems to conflict with the expectation that God will act in the same sort of way everywhere. One of the things many people have tried to learn in the modern world is that old assumptions about the inherent superiority of our own beliefs and practices need to be given up, as we understand more about other cultures. It seems much more tolerant to see our religion as one among others, and to see Jesus as one teacher or religious figure among others, rather than to assert that God acts uniquely in him. Why can we not see the way of Jesus as one way to God among many?

There is an important sense in which even (or especially) wholly orthodox Christians can and should do so. There are indeed many ways to God. God does act in all cultures to show something of the divine nature and purpose. The way we understand our own beliefs is very limited and fallible. These are important facts, which religious believers sometimes forget, largely because we are still prey to that greed, hatred and ignorance that Buddhists discern so accurately. Greed leads to the desire for power, which makes us see our religion as 'inerrantly true', whereas all others are to be mocked or suppressed. Hatred leads us to see members of other faiths as blasphemers or idolaters. Ignorance leads us to reject the beliefs of others when we do not even understand them properly. Since religion is bound up with human nature as it is, it will never be free of these tendencies. It is certainly important to try to counteract them, to treat the beliefs of others with respect, and to learn from them.

Nevertheless, it is not possible to believe everything at the same time. Disagreement is part of religion, too, and it is wrong to pretend that every belief can be equally true. What we have to learn is that divine revelation is received by fallible human beings, and if we are fallible, any of our beliefs may need correction. So we may have to disagree with others, or even with our past selves, to discover where this correction might be needed, though we should do so with respect wherever that is possible. There are many disagreements in religious belief. Some believe in reincarnation, whereas others think we are only born once on earth. Some think God is a personal agent, whereas others think the supreme reality does not act in the world. Some think the soul is immortal, while others think that there is no life beyond death. It is not possible for all these beliefs to be correct. So while we can certainly say that God is at work in all human cultures, we cannot say that all human beliefs about God's nature and purpose are more or less equally acceptable.

When we come to consider Christian claims about Jesus, we find that they are not just claims about whether some general teaching is more or less true. They are also claims about Jesus himself, claims that he really was in some sense the founder of the kingdom of God on earth. It looks as though Jesus was either mistaken about his central message, the coming of the kingdom, and the disciples were mistaken about his role in it as liberator and ruler of Israel, or he really does have a historically unique role to play in human history.

The Literalist Fallacy

It is quite possible to interpret Jesus as a Galilean exorcist, healer and preacher, who taught that God's kingdom was about to come, putting an end to the world as it was then known, and who died in an attempt to force God to bring history to a close. That was how Albert Schweitzer, a great biblical scholar, saw it in his book, *The Quest for the Historical Jesus*. But on that view Jesus was wrong in his central teaching. He must pass into history as one more failed preacher of the end of the world. It is hard to see such a man as one who should be taken as a supreme religious authority. It would be impossible for the Christian Church to take such a man to be the very incarnation of God.

There is another way to see the story of Jesus, however. Proponents of the alternative story might think that Schweitzer's view was conditioned by two main presuppositions, neither of which is obviously true. First, Schweitzer assumes that the miracle stories were all grossly exaggerated, especially the story of Jesus' resurrection. Schweitzer assumed that God does not act in the universe in specific ways, so a naturalistic account has to be given of the life of Jesus. He could be an exorcist, healer and preacher. He could even be a Messianic claimant, hoping that God would miraculously deliver Israel from the power of the Roman Empire. But he could not be a man in whom the wisdom and power of God was plainly manifest, who founded a Church to spread throughout the earth, and who was really raised from death to appear alive to his disciples. Schweitzer's first presupposition is that God does not act in historically unique ways for human salvation, and so could not have acted uniquely in Jesus. If, however, one accepts the possibility of divine action, it becomes plausible to think that God might have acted miraculously through Jesus (i.e. in ways transcending the natural powers of objects) to realise an important purpose on earth. Jesus may really have been raised to life after his death. Then it would not be enough to call him a healer and preacher. Jesus would be unique in human history, the only person who returned to the earth from the world of the dead.

Schweitzer's second presupposition is that religious beliefs are almost always to be taken literally. If Adam and Even took forbidden fruit in Eden, then there was a tree in some garden, and they physically ate its fruit. If Jesus preached the end of the age and Judgment, that means that angels will visibly appear in the sky blowing trumpets and all the dead will be raised out of their tombs. If Jesus preached hell for those who reject God, that means that their bodies will burn for ever if they have never heard the name of Jesus. If Jesus taught that he would soon return in glory, that means that a young Jewish man would collect his disciples into the air sometime not later than around 100 CE.

It is not hard to see why Schweitzer thought all these beliefs were false. Quite simply, Jesus did not return on time. This sort of literalism is fatal to Christian belief, because Jesus was not literally the Messiah, the deliverer and ruler of Israel. That is why Jews reject the Messianic claims of Jesus. He did not bring the twelve tribes back to Jerusalem in triumph and drive out the Roman occupation forces. For Jews, it is obvious that the

Messiah has not yet come. To make the claim that Jesus is the Messiah plausible Christians have to reinterpret the idea of the Messiah in a non-literal way. This is what they did, seeing Jesus not as a political deliverer of the nation of Israel, but as the spiritual deliverer and ruler of a new community of divine love, the Church, which would have its origin in history but would be fulfilled beyond historical time, at the consummation of all things. From the first, Christians adopted a non-literal – that is, a spiritual – interpretation of Jesus' Messianic role. It is because Schweitzer did not recognise this fact that he regarded Jesus as a failed prophet of the end of the world.

Total literalism, besides being fatal to Christianity, is in any case a complete misunderstanding of many of the spiritual truths the Christian faith teaches. I have already tried to show how the Genesis accounts of creation should be taken non-literally, if one is to understand the enduring spiritual truths they teach. The same goes for the teachings about the 'end of the world' and the coming of the kingdom of God that form the heart of Jesus' message. I shall give a short interpretation of these teachings later, in chapter 15. Two thousand years ago it was possible to take these teachings literally. Given the world-view of that age, many people probably accepted that the earth had only existed for a few thousand years, and that it might well end in a cosmic cataclysm in the near future. Within that world-view, part of the Christian message was that the end of the earth would not be the end of human existence, and it was not something to be feared. Whenever it happened, God would liberate those who trusted in the divine promises into a fuller life in a new creation. The kingdom established by Jesus would not be destroyed, but would achieve its full realisation with the ending of suffering and evil that the end of the earth would bring.

So in what is generally thought to be the earliest New Testament document, the first letter to Christians at Thessalonica, Paul exhorts the believers 'to wait for his Son from heaven, whom he raised from the dead, Jesus who delivers us from the wrath to come' (1 Thessalonians 1: 10). The 'wrath to come' is the envisaged ending of the earth in fire, which will be destruction and judgment to all who are evil, since it will put an end to the present world order. Paul's message is that God delivers from destruction and gives eternal life in a new creation. This promise was given in the life of Jesus, who was raised from death and continued to exist in a different, glorified form with God. The end of the earth, Paul taught, would also be the birth of a new creation, and its form would be the form of the transfigured humanity that the apostles saw in the risen Jesus.

The spiritual truth of this teaching is distinct from its literal interpretation, and already within the New Testament a different, literal worldview begins to emerge. In the Letter to the Romans, Paul writes that 'a hardening has come upon part of Israel, until the full number of the Gentiles come in, and so all Israel will be saved' (Romans 11: 25). As the Christian message began to spread in the Gentile world, it became clear that an early end of the earth would leave the vast majority of humans without the chance to hear about God's redeeming love. Soon, Christians began to pray that the earth would not end soon, so that the message of divine love could spread throughout the globe. The second letter of Peter makes the point explicit: 'with the Lord one day is as a thousand years, and a thousand years as one day' (2 Peter 3: 8). Christ will come again – that is, there will be a fulfilment of the kingdom, in which peace and justice reign and suffering and evil are destroyed for ever. In that kingdom, Jesus will be clearly revealed as the man in and through whom God acted in human history to deliver humans from selfish desire and make them 'partakers of the divine nature' (2 Peter 1: 4). But the time of that 'coming again' is in God's hands, and it will only be when the 'full number' of those who are to hear the gospel is complete.

Since the sixteenth century, our view of the universe has expanded enormously, and we now know that the end of the earth will not by any means be the end of the universe. It will only be a small blip in cosmic history. So for us a literal interpretation of the biblical accounts of creation and the end of all things is ruled out. It is important to see, then, that the literal account was never essential to understanding the deep spiritual truth of these biblical stories. Total literalism has now become the greatest obstacle there is to understanding the truth of Christianity.

The Sacred Cosmology of Christianity

Of course, there must be some literal truths in Christian teaching, or it would have no content. Jesus must have existed and been very much the sort of person the Gospels say he was. God must really exist and there must be a continuance of human life beyond death. But such truths are typically expressed by the use of highly symbolic imagery, especially when they deal with events (like the origin of the universe) so remote in history that no one can remember them, or (like the resurrection of the dead) so far beyond present human experience that we can hardly imagine what they will be like. The accounts of Jesus' life in the Gospels are not in this category, as the events were well within the memory of many people then living. Even then, those accounts would have been written so as to express specific spiritual meanings and insights that probably would not have existed at the time the remembered events occurred.

When it comes to questions of an overall view of the nature of the cosmos, it will not be surprising if the central spiritual teaching needs to be completely disentangled from contemporaneous beliefs about how the universe began or will end. We will naturally place the spiritual truths of the Christian message into the evolutionary form of our modern worldview. That worldview may turn out to be mistaken or quite inadequate at some future time, and yet we may be sure that the spiritual message will remain intact. I would think, however, that the general evolutionary view is so well established that it will not be completely overturned, even though it may be drastically modified. It might be as well to be agnostic about some of the finer points of modern scientific theory, but the overall view does seem fairly definitive. As I have argued, that general view actually helps to give a more plausible account of Christian faith, in many ways. But that will only be true if one has a firm grasp of the spiritual, non-literal sense of much scriptural narrative. It is because Schweitzer chose not to recognise that fact that he could regard early Christian teaching as false prediction, rather than as a profound teaching about God's kenotic, unitive and redemptive love, conveyed in symbolic forms.

So, for example, when Jesus teaches about hell, he is warning of the course of self-destruction upon which people embark when they choose lives of selfish desire. The fires of hell are the fires of unquenched desire, and the outer darkness is the separation from God that love of self inevitably produces. Selfish desire does bring suffering and destruction. In that sense hell is real, but it is not a literal fire from which there is no escape.

On the contrary, the whole point of Jesus' teaching is that people should repent – turn from selfish desire – and accept God's free gift of eternal life. When he speaks of 'the coming of the Son of Man in glory', he is speaking of the emergence of a truly personal community, filled with the joy of God's presence, when all evil and suffering has been overcome. That community will only emerge fully at the end of time, but within the lifetimes of those who heard him teach it can already be seen as a spiritual reality in time.

In other words, Jesus was not prophesying that the world would physically come to an end within a few decades at most. If he had been, Christianity should have died out, as a falsified millenarian sect, almost at once. Jesus was intending to found a new Israel, a 'kingdom not of this world' (John 18: 36), a community in which he would be permanently present as ruler. In his person, that kingdom was already present, for he is one who unites human and divine in himself. His humanity is freed from selfish desire, and is the vehicle of the rule of the divine Spirit. Knowing himself to be uniquely one with God, at once fulfilling and transcending Jewish Messianic hopes, he gathers a group of disciples, formed around 'the Twelve', leaders of the new Israel. He proclaims the birth of a new Spirit-led community, and prepares to send it throughout the world, where it will be the 'body of Christ', the community through which all things can be liberated from self and ruled by the Spirit of God. He promises his presence with that community, and teaches that there will be a consummation of all things, in which he (the cosmic reality of Christ, embodied in what will then be the glorified and transfigured humanity of Jesus) will be known in his true form.

This is, of course, one version of the picture of Christ that is held by the orthodox Christian churches. It is not some sort of radical new interpretation that exposes the orthodox tradition as a fraud or invented myth. Yet it is radical in at least two ways. It places the orthodox Christian story squarely within the modern scientific evolutionary worldview. This is clearly different from the worldview of biblical times, and yet the surprising thing is that, at least in my view, it gives even deeper meaning to the orthodox story.

It also insists on a non-literal interpretation of the 'sacred cosmology' of the Christian faith, the religious imagery of creation, fall, atonement, heaven and hell, the return of Christ and the Last Judgment. Many Christians will find this radical, because unfortunately total literalism has become so widespread in recent times. Such literalism is doomed, however, not only because science has shown that the universe is much bigger than literalist accounts suppose, but also because it misses the spiritual depth of Christian teaching about the cosmic purpose of God.

Albert Schweitzer thought that Jesus preached that the world would end in cataclysmic judgment within a few years, in which case Jesus was, to put it bluntly, just another crank. Suppose, however, that Jesus was a profound spiritual teacher, who taught in 'mysteries' and parables of a new reality, the community of the Spirit of God. In this community, humanity and divinity would be united, and all things would be reconciled in the cosmic Word. Jesus himself founded this community, which would be patterned on his life, and would indeed express the life of the cosmic Word that is one with his humanity, with the vocation to unite the estranged world to God.

The fact is that Christianity requires one to make a decision, not just about the truth of some general spiritual teaching, but about the person and role of Jesus of Nazareth. The Christian Church claims that he was indeed the inaugurator of the kingdom of God, a new spiritual reality existing, however ambiguously, in history, and to be fulfilled beyond history. If this claim is justified, then the belief that Jesus was the earthly incarnation of the cosmic Christ will not be a myth, almost comically exaggerating the importance of an obscure Galilean prophet. It will be a fact that marks Jesus out as unique in world history, since he is the human person in whom God acts to establish a community in which the divine purpose for this planet is to be declared, and through which it can be at least partly fulfilled.

This may seem like asking too much of the Christian Church. Surely it is not that much better than other religious communities? And is it not rather arrogant to claim that God's purpose is to be fulfilled through just this one religious community, rather than through many communities throughout the world? At this point the problem the Church has is very like the problem Jews have, about being a people 'chosen' by God. Both Jews and Christians have at times thought this 'chosenness' gave them some inherent superiority. But in their wiser moments, they have both seen that they are not chosen for their virtue or wisdom. The apostles were neither wise nor virtuous. They often seem particularly obtuse, power-hungry and fearful. It is better to see the Church as a community of forgiven sinners than as a society of perfect saints. The Church will not be 'better' than other social groups, and it should not base its claims to truth on such a fragile base.

One claim that is central to Christianity is that Jesus is a servant-Lord. By his direct command, his followers are to serve the world as he did, not dominate it (John 13: 14 and 15). So the only claim the Christian Church should make about itself is that God has called it, through Jesus, to serve the needs of the earth and work for the establishing of a global society of justice and mercy. There are many other communities in which God might be known and served, and Christians should co-operate with them in working for the welfare of the world. Christians should rejoice with them in the knowledge and service of the creator, and learn from them what God has revealed in their distinctive paths. Christians should not claim that God only works to fulfil the divine purpose through the Church, or even some favoured part of the universal Church.

The distinctive path of Christianity is that it should form a community in which self-giving divine love can be mediated for the welfare of the earth, and in which human lives can be liberated from selfish desire, empowered by the divine Spirit and fulfilled by being made sharers in the divine nature. Of course, Christians believe that when God's kingdom fully comes, it will be the fulfilment of what is so ambiguously and fitfully present in the Church – just as Buddhists believe that the final human goal is nirvana, and Muslims believe that it will be the resurrection to paradise of which the Qur'an speaks. Perhaps the final human goal is not less than, but so much more than all that our religious images convey that many different images will find their fulfilment there, in unimaginable ways. For the present, it is surely enough for Christians simply to witness to the love of God, which has been made known to them through the person of Jesus, whom they believe to be the Christ, the liberator and ruler of a community chosen to be a vehicle of divine love on this planet. That is the calling of the Church, and that is the criterion by which the 'true Church' should be judged.

13

Christianity Among the World Religions

The Religious Traditions of Humanity

On earth, the existence of the kingdom of God in its fullness has been made impossible by the fact that humans are trapped in desire, hatred and greed, ignorant of God and subject to despair, suffering and death. What God wills is that human persons, like any and all finite rational beings, should live in fully conscious knowledge of God, as perfect channels of the divine wisdom, power and love. Human persons are meant to be vehicles for the manifestation of the divine love, sacraments of the divine life, temporal images of the eternal reality of God. The Book of Genesis, the first book of the Bible, states that humans are 'made in the image of God' (Genesis 1: 27). A human life should be a finite picture, or image, of the nature of God, and a mediator of divine creative power.

Human beings have defaced, but not perhaps completely lost, that image in their own lives. At the very beginning of human history, when conscious beings first became faced with the challenge of divine justice and the promise of divine love, they turned aside from the path God wished them to follow. They chose the path of selfish desire and freedom from divine control. We, their descendants, are now trapped in a world of desire, greed and hatred, in bondage to the illusory self, estranged from the presence of God. We are not what God intended us to be. We are, in that sense, less than truly human. This is the 'fall', better thought of, perhaps, as a 'failure to rise' to the mature, loving and harmonious human community God had intended for us. Its consequence is that all humans are now born in what has been traditionally called 'original sin' – unable to escape completely from that greed, hatred and ignorance which lies in our evolutionary inheritance, and which society nurtures with all its power.

Desperate situations require desperate remedies, and if we are aware of our plight, we might hope that God would somehow act within our world, without destroying the freedom God has given but nevertheless breaking the power of evil and reuniting human lives to the divine life. We might hope for liberation from bondage to self, by a form of divine action that is not dictatorial and yet is effective in uniting us to God. Throughout human history, we might hope that there would be many ways in which God begins to liberate humans from evil and reunite them to the divine. The forms divine action takes, and in which it is understood, will vary with diverse cultures and histories. Sometimes human understanding of the divine will be almost overlaid by the distorting factors of fear and hatred, so that violent and intolerant forms of religion will dominate a whole society. Sometimes a rather narrowly tribal or ethnic interest will predominate, so that local symbolic forms and practices will resist the development of a more global view.

The history of religions is partly a history of human ignorance and folly, and partly a story of the development of more adequate understandings of God and the way to the human goal, in response to new factual knowledge and wider human sensitivity. Many students of the history of religions point to three main streams in which such development has taken place. One is the Semitic stream, which begins with Hebrew belief in a tribal God who liberates from oppression, and develops a prophetic tradition of judgment on injustice and liberation into a truly just and merciful society. In this stream, the idea of God as a moral authority and transformer of history becomes the dominant image of the Hebrew Bible. The proper human goal is seen as the establishing of a society of justice and mercy, in which individuals can fulfil their distinctive personalities in relationship one with another.

The Indian stream develops in a different way, from rituals of sacrifice to the gods and spirits of nature to the idea of one supreme reality of wisdom and bliss which diversifies itself into the finite universe, unity with which can be realised by withdrawing the mind from the senses. In this stream, the idea of Brahman as the inner reality of all things, to be known by the renunciation of action and desire, becomes the dominant image of the Upanishads. The universe is under the sway of karmic law, and the dominant religious goal is to obtain release from karma, never to be reborn.

The Eastern stream, in which Buddhism, Taoism and Confucianism interact, develops from forms of animism to the idea of a cosmic order, a way of balance and harmony, following which brings stability and calm of mind, and peace and right order in society. In this stream, there is little stress on one absolute being or God. Emphasis is placed on living in the unending flux of beings, without attachment but with mindfulness and compassion for all suffering beings. The goal is to leave behind any idea of 'self' or subject–object duality, and to experience the vibrant flow of being, beyond passion and attachment.

These characterisations are of course over-generalised, and they should not be regarded as inflexible and unchanging formulae which forever separate these traditions. Nevertheless, they do give a general idea of how religious perceptions have developed in different ways in different cultures. There is no reason why a theist should not see God as at work to reveal and liberate in all three streams. Obviously, however, there is little sense of a personal and active God in the Eastern stream, and not much sense of a God who is interested in individuality and the particular processes of history in the Indian stream. To that extent, any form of theism that stresses hope for an eventual realisation of a just society, the importance of individual personality, and the working-out of a moral purpose in history under the guidance of a providential God will belong to the Semitic religious stream.

In our age, when we are able for the first time to think globally about religion, it may be appropriate to seek some convergence of these great traditions. The regrettable intolerance and violence that so often marks the Semitic tradition may be mitigated by the Asian stress on compassion for all beings and equanimity of mind. Indian stress on the unity of all things and on the spiritual path of non-attachment may be complemented by Semitic insistence on the importance of social justice and the dignity of individual persons. The possibilities for fruitful interaction are immense, and indeed there are many people today who in their own lives combine various elements of these traditions. A meeting of two or more great complementary traditions may enable some of one's own cultural restrictions to become clear, and may transform the understanding of each tradition in a wider, though to be sure still limited, perspective.

For example, if Christianity interacts with the Indian tradition of Vedanta, it may provide a distinctive interpretation of the Indian tradition, by stressing the way in which God enters into a particular history to act in new and creative ways. It will stress the particularity and the importance of events in history in a way that the vast cosmological perspectives of Vedanta may overlook. It will stress the absolute moral demands for justice and compassion that God makes, but also the suffering and serving form of the love of God, which will forgive and reconcile those who confess their failure to meet those demands and turn to God for help. It will stress the features of personal relationship and community, which do sometimes get overlaid in Vedanta by more introspective and individualistic patterns of spirituality.

On the other side, an understanding of Vedanta can help Christians to see that God is not only the severe Judge who condemns humans as miserable sinners, but the Self of all, with whom all sentient beings are united indivisibly at the very heart of their being. It can help Christians to see that there are many paths to God, and that those paths must be evaluated by their capacity to lead to non-attachment, wisdom and compassion, not simply by their conformity to some doctrinally correct formula. And it can help Christians to see how persons are pursuing a spiritual path in which each is responsible for their own liberation or bondage, under the forgiving and guiding love of God.

In the third millennium many possibilities exist for bringing the various religious traditions into a positive and mutually enriching relationship. For that to happen, the traditions must not be destroyed, but must remain as witnesses to the diversity of human understanding of God, a diversity which will remain within any wider convergence of traditions. Within this diversity, the Semitic tradition will properly continue to maintain its own distinctive witness to the revelation to the prophets that God is moral will and personal Lord, even while it seeks to extend its vision by reflecting from its own viewpoint the insights of other traditions. It can bear witness to a non-exclusive, non-exhaustive truth, imperfectly understood, which nevertheless describes the nature of God, the final human goal and the way to the goal in a way that seems adequate to present possibilities of human understanding.

The Distinctive Christian Tradition

Naturally, the Semitic tradition is not monolithic. Within it there are many diverse perspectives, which develop as humans reflect on the revelations they believe the prophets to have received. Orthodox Judaism remains loyal to the revelation believed to have been given to Moses in the written and oral Torah. Orthodox Jews feel that God has chosen Israel to be in a particular covenant relationship with God, so that they are separate from all nations on earth, to follow God's revealed law for ever. Less orthodox (Liberal, Conservative and Reform) forms of Judaism agree that Jews are called to witness to God's glory, justice and loving-kindness (*chesed*) in a special way, but tend to give Torah new interpretations in different cultures, and stress the universal principles of social justice and personal integrity rather than some of the more particular rules to be found in the Babylonian Talmud.

Islam accepts the Semitic insight that God reveals laws for social justice, but the Qur'an issues a universal law (*shariah*) for the whole world, or at least for a global *ummah*, or community of Islam. Again, there exist many varieties of Islam, but they all agree in believing that Muhammad received a definitive divine revelation, the Qur'an, which lays down the principles and the goal for a truly just society and promises life in the world to come for those who obey the commands of a compassionate and merciful creator.

Within the Semitic tradition, Christianity developed in a different way. It began as a Messianic school within Judaism. Many of the prophets had looked for a Messiah, a great liberator and ruler who would deliver Israel from subjection to the great world empires, and enable Torah to be lived out fully in a secure and peaceful community. There were many ways of seeing the Messiah, ranging from the idea of a straightforward political leader, a king in the line of David, to a supernatural 'Son of Man', who would put an end to the unjust world order by a decisive divine intervention.

What was distinctive about the Christians is that they believed Jesus of Nazareth, a healer, exorcist and preacher from the rural province of Galilee, to be the Messiah. He was obviously not a political leader, and he did not liberate Israel from Roman occupation. In fact, he was executed as a criminal or troublemaker, and not long after his death the nation of Israel was completely destroyed by the Romans. Jesus was a different kind of Messiah, one whose kingdom was not of this world. His disciples believed that he liberated them from the bondage of selfish desire and made possible a new and intimate relationship to God as they followed him. They believed that they experienced his presence with them after his death, and they were filled with a new energy and joy by the Spirit that he had promised to send them. They found, in their new community, living in the Spirit of Jesus, an image or foretaste of what God's kingdom was like. They believed that Jesus had been raised to the presence of the Father, vindicating his Messianic claims and his promise that the kingdom of God would draw near to the world in a new way. They experienced that kingdom in their own lives, and hoped for the full realisation of the kingdom when human history came to an end, and all things would be united in the cosmic liberator and ruler, the Christ, whom they had seen in the person of Jesus.

The belief that Jesus was the Messiah put them outside the bounds of traditional Judaism, which still looked for a redemption of the nation of Israel. It gave birth to a new worldwide community, which they saw as the 'body of Christ', called to serve and bring reconciliation and forgiveness to the world, to proclaim in word and deed the self-giving love of God and to unite human lives to the life of God, which they had seen in Jesus.

Christians see their faith as a continuation of one strand of the Jewish tradition, now opened up to the Gentiles as a global community of love, which reveals the divine nature as kenotic (self-emptying) love, and sees the human goal as unity with the divine nature, a community 'in Christ', formed through the creative power of the Spirit. This is the distinctive witness of Christian faith, its contribution to the range of images of and paths to the divine in human history.

Finite Forms of God

The Christian reinterpretation of the Messiah as one in whom the kingdom of God is already realised expresses the possibility that if ever there could be a truly human person, intimately related to God, a true channel of divine love, in that person the distorted image of God could be healed and corrected and made clear and visible again. In this world, such a person might look more than human, supernatural or abnormal. But that person would, in fact, be a truly human being in a world of partial humans, sick souls estranged from the source of their being. And that truly human person could perhaps become a means by which the human world becomes able to return to God.

The early followers of Jesus certainly saw him as a person in whom the image of God is renewed and perfected, through whom the human world can find liberation from bondage to selfish desire, and by whom a new community of the divine love is founded. Jesus is referred to in the New Testament as 'the visible image of the invisible God' (Colossians 1: 15). He came, he said, to give his life so that many might return to God. If these things are true, Jesus is one through whom God acts to reconcile the world to the divine being.

In the Gospel of John (10: 38), Jesus makes an extraordinary claim to unity with the divine: 'the Father is in me and I am in the Father'. The claim is that the human person of Jesus is not just a human being whose life is similar to that of God the Father, the creator of all things. Somehow, his life is interwoven with that of the Father, so that they cannot in the final analysis be separated. We might say that the human life of Jesus is the form the divine life takes within the limitations of human history.

As one reflects on the simple truth that God is revealed in the life of Jesus, one begins to see the profound mystery at the heart of the Christian faith, the *mysterium Christi*, the mystery of Christ. For Jesus is not just someone who tells us what God is like, or who shows us by his actions what God is like. Jesus is someone whose life is the historical form of God. Here is perhaps the most startling claim that orthodox Christianity makes – that God has a historical form, and that form is a particular life in the history of the earth.

For monotheists in the Abrahamic tradition, it may seem shocking to say that God has a finite form. It is not, in fact, quite as foreign to that tradition as may seem from the strong prohibition on making any finite depictions of God, or on associating anything finite with God. There are many instances in the Hebrew Bible when God appears in human form, whether walking in the garden of Eden, wrestling with Jacob, or enthroned on the holy mountain. These are appearances of God in finite form, though, of course, they are not to be identified with the divine being in itself, which is beyond all representation. There are also well-known passages in the Qur'an which refer to the 'face' of God (2: 272), and to God as 'sitting upon a throne' (2: 255). Most Muslim commentators very reasonably take these as metaphors. Even so, they must be regarded as divinely given images that represent in a limited way something of the reality of God. It might even be said that if one could not use finite images, one could not think about God at all. One might have to be very careful to say that these images could not be literally ascribed to God, yet they may depict God in a way that is accessible to human minds.

There are also traditions in both Jewish and Muslim thought that come remarkably close to the possibility of divine incarnation. Some rabbis were spoken of as 'embodiments' of Torah, the divine teaching. If Torah is reified, as it sometimes was, to be the Divine Wisdom, it was possible to speak of some humans as incarnations of the Divine Wisdom. In Islam, too, the doctrine of the unity, sovereignty and omnipresence of God can be so emphasised that all things become immediate expressions of the divine will, which is one with the divine nature. Surprisingly, in the pure monotheism of Islam there have been many mystics who claim to have experienced the fading away of their own personalities and wills (*fana*), leaving only the reality of God, of which they are immediate expressions.

Though such ideas are on the fringes of the Abrahamic traditions, they represent possible developments of those strictly monotheistic traditions. It seems possible, even in those traditions, for the infinite God to have a finite form, or to be expressed in a human life. In Indian religious traditions such a possibility hardly seems unusual. Vishnu is said to come to earth from time to time as an avatar, a finite appearance of

God, and many Indian gurus are said by their followers to be earthly appearances of the supreme reality.

Christians do not, however, think that Jesus was a human appearance of God. He was a real human being, not just someone who seemed to be a human being but was not. As a human being, he possessed a human mind and will, thinking, experiencing and acting largely in response to the information conveyed by his senses, like any other human being. He was not a human body with a divine mind. He was a human mind and body, which was truly the historical form of God.

How can a human mind be the form of the mind of God? If we say that a human body is a finite form of God, we do not mean that God really has a human body, even when there is no universe. We mean that God can express the divine nature for the sake of embodied beings in the form of a body, which they can see and respond to. The body is a temporary form, for the sake of devotees, and its function is to point beyond itself to the nature of the divine, which it in some way appropriately symbolises.

In a similar way, God is not essentially like a human mind. A human mind can, nevertheless, express the divine nature for the sake of embodied beings, always pointing beyond itself to the divine, which it symbolises. It can do so only if it freely and consciously attends at all times solely to God, and submits itself completely to be used by God to express the divine nature, to be a finite symbol of the divine for the sake of embodied beings. Under those conditions, however, we might be able to say that God has become incarnate in a human person, enabling that person to express the divine nature and purpose in his or her life, and to mediate the divine presence and power through that human nature. Such a person might be able to break the power of greed, hatred and delusion in human existence, and open a path from the bondage of selfish desire to the freedom of the love of God. Christians believe that such a person would be the true Messiah, the Christ, and they believe that Jesus of Nazareth is the Christ.

The Incarnation

God could in theory take many minds and bodies to be finite forms of the divine nature. There is nothing to prevent the infinite God from taking any number of finite forms. But two main things are necessary if a human person is to be a finite form of God. That person must be wholly obedient to God, and there must exist a historical and cultural context which makes the expression of the divine nature in that person intelligible.

The number of people who are wholly obedient to God must be very small indeed. For such obedience to be possible, one must have a clear awareness of what God is and of what God wills – much clearer than most people have. One must have complete control over the selfish passions that form such a great part of most human lives. And one must have an unshakeable resolve to do whatever God requires – without the vacillation and compromise that characterises most human conduct.

A human life could have these qualities. Indeed, perhaps we may all hope to have them in that perfected form of human existence that we call 'heaven'. But all will agree that they are very rare indeed on this earth. A human life that shows perfect knowledge and love of God, and which is selflessly devoted to the service of God, is so rare that its appearance is almost, if not quite, a miracle.

It becomes truly a miracle if that human soul is such that it could not fall away from God, but is indissolubly united to the divine will. By a 'miracle', I do not mean an arbitrary interfering with laws of nature, but a fulfilment of human nature that enables it to transcend its normal modes of operation, by its deeper relation to the spiritual basis of all physical reality. In Jesus' case, the orthodox Christian view is that God united a human personality to the divine in such a way that Jesus had a virtually overwhelming sense of the presence of God and an unreserved openness to the love of God. Because of that overwhelming knowledge and love of God, he could not fall away from God. We can then say that his human existence was not distinct from the reality of God. In him, human and divine found a profound unity.

It is in that point that the human nature of Jesus is said by orthodox Christians to be different from all other human natures. It is precisely a human nature that is not distinct from God, and so it is not able to fall into selfishness, which is alienation from God. Jesus is fully human, except for the ability to sin – 'one who in every respect has been tempted as we are, yet without sin' (Hebrews 4: 15). To tempt is to try to attract to do something wrong through the prospect of some advantage. Being human, Jesus would have felt the attractions of selfish desire. But such temptations were bound to fail, since his love for God was such that it outweighed all other attractions. That, at least, is the orthodox Christian understanding of Jesus and what makes him unique among human beings. It means that Jesus is human, but is not distinct in existence from God. He is God in human form, with a true human nature. To believe that is to believe in incarnation, that the eternal Word of God, uncreated and undying, assumed human nature, that 'the Word became flesh and dwelt among us . . . and we beheld his glory' (John 1: 14).

That seems to be a coherent possibility, though obviously other human beings will never, even in heaven, have quite the same relationship to God as Jesus. We will always be individuals distinct from God, though we hope we shall be wholly united to God by the power of divine love, which we will have freely and gladly accepted.

Jesus is, of course, free in many ways – free to choose to do one thing rather than another, free to direct his energies in many different directions. But he will never exercise this freedom by disobeying God. One might say that the grace of God so suffuses his human nature that it unites it from the first moment of his being, and indissolubly, to the divine. He is, by nature (not by choice), what we hope to become, by what Paul calls 'adoption' (by free acceptance of the grace of God). In this respect, Jesus is said by the orthodox to be unique among human beings.

Living with Disagreement

This idea of a real human nature united in the closest way to God is not found in the prophetic religions of Judaism and Islam, where prophets always remain distinct from God. And it is not found in the Indian Vaishnava traditions, where the avatars of Vishnu do not suffer the limitations of having a real human nature. In some ways, Gautama the Buddha is more like Jesus, because (at least in many forms of Buddhism) he is a human being who has so perfected his nature that he transcends the human and becomes identical with ultimate reality. But the Buddha is not and has never, by Buddhists, been claimed to be an incarnation of God. Indeed, classical Buddhism has no belief in one creator God. Gautama is said to have attained perfection by his own efforts over many lifetimes. Jesus, however, is believed by Christians to possess a human nature that is perfected by the grace of God, and his loving relation to the Father is rather different from the Buddha's attainment of the perfect calm of nirvana, which seems to have transcended all relationships.

So, though there could in principle be many incarnations of God, there is only one realistic historical claim to such an incarnation, and that is the claim made for Jesus. The claim becomes plausible in a historical and social context which makes intelligible the idea of a human person in and through whom God acts in a paradigmatically revealing and salvific way. The long history of ancient Israel and Judah, with its central prophetic belief in saving acts of God in history, its expectation of a Messianic saviour, and its tradition that humans are created in the image of God, provides such a context. Within that unique history, it became possible to see a particular human life as a perfect image of God and a focal embodiment of God's liberating action to unite human lives to the divine. In that historical context, Jesus can be seen as the earthly incarnation of the divine nature and the paradigm case of divine redemptive action.

This does mean a disagreement with Jews, Muslims and all who do not find it possible or plausible to speak of a divine incarnation. But we have to face the fact that there will always be such disagreements. Giving up belief in the incarnation will not magically make disagreements about religious doctrines disappear. Orthodox Jews do not believe that Jesus was of greater authority than Moses, and regard the cavalier attitude towards Torah that his followers display as wholly retrograde. Orthodox Muslims generally deny that Jesus died on the cross, and regard the teaching of the Sermon on the Mount as unrealistic and probably harmful. We are not going to eliminate disagreements by denying the incarnation.

Some people are very unhappy about the existence of disagreements between religions, and try to devise forms of belief that would eliminate them in some way. I think one has to say quite simply that this cannot be done. Every such proposal – and one can think of Wilfred Cantwell Smith and John Hick as eminent Christian thinkers who have tried such a strategy – simply produces more disagreement! The only way one could do it would be to drop all beliefs that religions disagree about, or reformulate them in ways that could be agreed, but that is impossible. Suppose a Christian believes in a personal creator and a Buddhist does not. One might suggest dropping the idea of God – but that would hardly satisfy the Christian. Or one might suggest getting the Buddhist to accept some rather vague idea of God, such as 'Pure Mind'. But Theravada Buddhists have long ago rejected such Mahayana Buddhist ideas, so there is not much hope of progress towards agreement there.

The next move might be to admit that mainstream believers are not going to give up beliefs they regard as central – such as incarnation, in the Christian case. But one might suggest that one simply leaves orthodox believers alone in their benighted ignorance, and recommend revisions for the enlightened, which they could all agree on – note how elitist and patronising such a view would be! What might one end up with? One possibility is a sort of 'highest common factor' view, present in nearly all religions, that one should overcome selfish desire, and find some sort of unity or conscious relationship with a reality or state of wisdom, compassion and bliss (this would allow Buddhists as well as theists to agree).

I think this is a very attractive proposal. But it will still not get universal agreement, even among the 'enlightened' liberals, because large numbers of liberal people will deny that there is just one such state or reality that is a possible and realistic goal for human beings. Some may even angrily reject the idea of aiming at such an allegedly universal goal. Also, as soon as one tries to spell out what the goal is, how it is to be attained, and who is in the best position to know such things anyway, new disagreements will be generated. The empirical evidence of this is the large number of new religious movements in the modern world, many of them claiming that 'all religions are true', which nevertheless form new religions, sometimes persecuted by older religions, and which all have distinct particular beliefs.

Universal agreement among human beings about anything is an impossible dream, and this includes religion. It is hardly surprising that there are disagreements in an area where human capacities are so limited, human cultures and histories are so diverse, and the objects of discussion (ultimate reality and the final destiny of human beings) are so far beyond adequate human comprehension. Recognition of these points should lead to a great degree of tolerance of diversity in religion, and might also suggest a good deal more agnosticism, or at least humility, than one often finds, but it makes disagreements more, not less, likely.

It is entirely reasonable to celebrate disagreement as a safeguard against complacency in error, and a spur to seeking new insights into truth. Perhaps the optimal aim would be to ensure that religious belief and practice produces joy, peace, justice, compassion and kindness, and that we never let theoretical disagreements stand in the way of achieving those fruits. Most religious teachers have aspired to encourage such qualities, though I suspect that, at least in the West, it is actually humanism that has persuaded religions to place them as high on the agenda as they should always have been.

There is, I think, very little hope that Christians will come to agree on all religious doctrines either with members of non-Christian faiths or with each other. The very liberal Christians who most desire that end are precisely those who most disagree with most of their fellow believers. But if I think that a specific belief – for instance, that God was uniquely incarnate in Jesus – is true, does that mean I must think my view superior to all others, or that I must regard different beliefs as alright for people of lower intelligence or insight than myself? One must be very careful here. It should be obvious that people of equal intelligence and sensitivity hold very different religious and non-religious beliefs; so I should never say that my holding specific beliefs makes me superior to other people. On the other hand, it is equally obvious that I must hold that a true belief is in some sense superior to a false belief. It is superior simply in the sense that I should reject the false one in favour of the true one. Of course, it does not follow from this that I am infallible or in a specially privileged position to know the truth. It is just that I cannot hold both of two conflicting beliefs at the same time. A Christian must think that the belief that Jesus died on the cross is 'superior' to the usual Muslim belief that he did not. But equally the Muslim will think that her belief on this matter is 'superior' to the Christian one. Both can agree that neither is stupid or depraved, and that both are fallible and could be mistaken.

People should not simply rest content with disagreement. It is important to try to find out why we disagree, what are the true grounds of our disagreement, and to what extent we may be able to move to a wider view that might indeed resolve it.

There will still be many points, however, at which disagreement with others remains. Anyone who feels uneasy about this might reflect that, if they disagree with what I am saying, they are committed to affirming that at least one pretty fundamental disagreement between religious believers exists! That does not mean Christians (or anyone) should hate or despise those with whom they disagree. Christians are commanded to love even their enemies. And those who disagree certainly do not have to be enemies. It is a mark of maturity to be able to disagree in friendship, and that is what religious believers in the pluralist context of the modern world are going to have to do. What this brief discussion of a very complex and pressing issue suggests is the need for a firm commitment to what might be called seven principles of global pluralism:

- We must respect and encourage the freedom of others to form their own beliefs conscientiously.
- We must affirm that people are not to be condemned or blamed simply for their (or our) intellectual errors.
- We must resolve always to seek the welfare of all, without distinction, whatever their beliefs.
- We must seek positively to learn from the differing viewpoints of others, even though we cannot agree with all of them.
- We must seek a convergence of views wherever possible, since truth is one. We must try to discern complementarities, eliminate stereotypes of others, and be sensitive to the cultural and historical influences on our understanding.
- We must co-operate for good with all who are committed to the pursuit of truth, beauty and goodness, however different the paths we follow.
- We must try to ensure that our own religious beliefs lead to reconciliation and peace, to personal well-being, mindfulness and compassion, and not to violence, anger or censoriousness.

Religious people, and non-religious people, too, often find it very difficult to live up to these principles. But there is no reason why believers in a unique incarnation of God in Jesus should not accept them – and every reason why they should, because Jesus teaches that God desires the salvation, the ultimate welfare, of absolutely everyone.

Jesus and God's Universal Love

It is important to remember that those of us who believe that God became incarnate in Jesus do not think that the creator turned into a man. That is why we distinguish God the Father, who always remains the creator and sustainer of the whole universe, from God the Son, the Word, who 'became flesh'. But even the Word does not *turn into* a human being. Rather, a particular human nature was assumed by, uniquely united to, God the eternal Word. Jesus becomes the true historical embodiment of the Word, but the full reality of that Word far transcends any finite form. The Word is the cosmic Christ, through and in whom the whole universe was created, and in whom that universe will find its fulfilment. When we say that God the Word was incarnate, we mean that the life of Jesus is the historical form on earth of that cosmic and trans-cosmic divine reality.

Another metaphorical way of putting the same point is to say that Jesus is the 'only son' of God. That does not mean that Jesus is somehow physically or genetically related to God – an absurd supposition. It means that Jesus has a unique form of relationship to God, a relationship so close that an apt metaphor to describe it is the relation of a loving son to a loving father. The relationship is unique, because this life, in its unrepeatable historical context, becomes the place where the eternal Son, the Word, is embodied and expressed in the human world.

Through that incarnation, a human life achieves its true destiny of becoming the perfect channel of the divine life. Thereby, humanity is united to the divine, and God's cosmic purpose is fulfilled in one life. But the divine purpose is that all humans should be able to share in that fulfilment. The life and teaching of Jesus is meant to open up a way for the whole world to find that release from selfish desire and complete union with God that was realised in Jesus. It is in opening such a way that Jesus becomes the saviour of the world.

It is unrealistic to think that the Christian Church this side of eternity will either embody this ideal perfectly or that it will ever become the one universal religion. Dreams of a Christian world, like dreams of a Muslim or Jewish world, have been ended by the realisation that we will live in a pluralistic world for the foreseeable future. Christians should be clear, however, that the universal salvation of all – if they do not finally reject God – through their coming to share in Christ and the Spirit, does not depend on the Church becoming universally dominant, as though one could only be saved within the Church. Such a view is not sustainable, since even if the Church became the one universal religion, most humans who have lived on earth would not have been members of it. All that is required is that the distinctive witness of the Church turns out to be true. For the distinctive and particular witness of the Church is precisely that God's love is unrestrictedly universal, that it is truly seen in the healing, forgiving and compassionate life of Jesus, and that it is truly (but not exclusively) given in the Spirit. In all probability, the Church will continue to exist as one religious community among others, one of a family of faiths, each of which holds in trust a distinctive insight into the ultimate reality and the final human goal.

Of course the Church will hold that its sacred cosmology of creation, incarnation, atonement and the reconciliation of all things in Christ is true. It is a cosmology that coheres well with the scientific cosmology of an evolutionary universe, structured to produce responsible persons capable of forming communities of creative freedom, sensitive knowledge and compassionate co-operation. It goes beyond the scientific picture in speaking of a spiritual reality that generates, enters into and transfigures the material universe into an expression and vehicle of its own being and action. It gives a further level of meaning to the scientific picture, as persons on this planet are seen, not just as products of the material universe, but as capable of transcending the material to find fulfilment in a trans-historical form of being, in so far as they learn to overcome selfish desire and share in the divine love.

If the Christian sacred cosmology is in fact true, all who are eventually reconciled to God will be so reconciled in Christ, through the power of the Spirit. Since most people do not now accept this, the truth of the Christian account entails that beyond this earthly life there will be opportunities to learn and develop and come to know God more clearly. This will be true of Christians as well as of others, so that we may be sure that our particular views of Christian truth may change considerably as we come face to face with truth. At that point, we may understand more fully what the importance is of the existence of so many different worldviews, and to what extent there can be a convergence towards a presently hidden truth between them. For the present, all one can do is to be as self-critical and open to as many visions of truth as one can. In that spirit, what I am trying to present is a mainstream Christian view in the light of contemporary scientific understanding and the need to adopt a fully global perspective. Seen thus, I believe that Christian faith gives a coherent and plausible account of how the world really is, though any decisions about that will largely depend upon the sorts of affirmative and liberating experience that Christian communities make possible for particular individuals.

14

The Life of Jesus

The 'Virgin Birth'

Many books have been written about the life of Jesus, some of them so speculative as to be almost unbelievable. The fact is, however, that the four Gospels are the only reliable records of his life that we have. They were collected together by the Church, by people who believed that he was the Messiah, saviour of the world. They selected the Gospels from a group of available materials, rejecting other, 'Apocryphal', works because they were considered to be too fanciful or unreliable. They collected them for the purpose of inspiring prayerful meditation and giving insight into God's will, not because they were interesting pieces of biography. What the Church wanted to do was to evoke in others the disclosure of God's love that the person of Jesus had evoked in the first disciples. So the only information we have about Jesus was collected by disciples for the purpose of religious teaching. It is not disinterested or critical history.

It follows that attitudes to the Gospels will differ, according to whether one is open to the idea that there may be a divine cosmic purpose, which was revealed in the life of Jesus, or whether one regards that idea as a fantasy. It is entirely reasonable for those who are open to the idea of divine revelation to suppose that the Gospels give a highly reliable account of the historical Jesus. But of course anyone who has studied religion will expect that the accounts will not just be straightforward histories. They will probably, like most religious narratives, set the story in a cosmic context, so tending to heighten the narrative and fill it out with material of a primarily symbolic nature. They will insert into the narrative religious insights that in fact only became explicit at a later date, and they will conflate the experiences of the living community of the Church with the first experiences of the disciples of Jesus. So one will probably not have a strictly literal history, but rather a symbolically heightened narrative that attempts to portray deep spiritual meanings in an objective physical form.

It is fascinating to see how each of the four Gospels presents its material about Jesus in a different theological context, giving us a multi-faceted portrait, different aspects of which may speak to us at different times. The first three, called the 'Synoptic' Gospels, present Jesus as a teacher of striking authority who announces that God's long-awaited kingdom is about to dawn, portrays the nature of the kingdom in vivid but cryptic parables, heals the sick and exorcises demons, consorts with notorious sinners and outcasts, forgives sins and gets into increasing trouble with the religious authorities. Although he does things that constitute claims to Messianic authority (like riding into Jerusalem on a donkey, in fulfilment of prophecy, appointing twelve followers to be leaders of the twelve tribes, and instituting a 'new covenant' with his disciples), he also refuses to make such claims publicly, and his followers constantly misunderstand him.

The Gospel of John, on the other hand, shows Jesus as delivering long speeches in which he explicitly claims to be the Son of God, in a unique sense. The parables of the kingdom are not recorded, and indeed the kingdom of God is rarely mentioned. Instead, there is a much greater emphasis on the inner union of the disciples with their Lord, and on the unique unity of Jesus and the creator. John's Gospel seems to be a profound theological reflection on the cosmic role of Jesus, which brings out aspects of that role that were largely implicit in the Synoptic Gospels.

Taking the Gospels together, one can see how Jesus transformed the expectation that a Messianic king would expel the Roman occupation forces and establish an independent political state to which all Jewish exiles could return. In its place, he interpreted the idea of a Messiah in terms of the 'Suffering Servant' of Isaiah 53, one who would suffer for the sake of his people, and give his life to bring about God's purpose. He interpreted the kingdom not as a reunited Israel, but as a new universal community whose members would share in the life of the cosmic Christ, actually becoming the body of Christ on earth, thereby continuing the incarnation of the eternal Word that had been decisively expressed in Jesus. The Synoptic Gospels record the way in which the apostles were slowly taught to understand this. John's Gospel records the fruit of the fuller understanding that dawned only after the resurrection of Jesus.

Some features are central to all the Gospel accounts. Even if he was said to be descended from King David, Jesus was not born into a wealthy or important family but into a fairly obscure family in a rural province. The birth narratives, which occur only in Matthew and Luke, are in stylised literary forms that show direct dependence on passages of the Hebrew Bible. Most biblical scholars regard the stories of the wise men and shepherds as legendary material of the sort found in many religious traditions (for example, in the lives of the Buddha and of Krishna) which see their founders as having miraculous births.

One can trace a development of such legendary material in the story of the wise men, an undisclosed number of magi, astrologers or priests who, over centuries, turned into three kings, and even had names and nationalities attached to them. Similarly, the 'virgin birth', Mary's conceiving of Jesus without fertilisation, later became a miraculous birth that did not rupture the hymen, and Mary became a perpetual virgin, who never had children, despite the fact that Jesus' brothers are referred to in the Gospels. As is well known, the prophecy that the Messiah would be born of a virgin is also based on a Greek translation of the Hebrew, which in the original prophecy just meant 'a young woman'.

In the light of these factors, many scholars hold that the traditional belief that Jesus was born of a virgin mother is a legend, a symbolic insertion into the story of his life. The symbolic significance is weighty. Israel is referred to in the Hebrew Bible as 'the virgin Israel', the bride of God (Amos 5: 2). Jesus is the first of a new humanity, mediators of the love of God, and he is born of a purified Israel, not by the will or action of any man, but by the pure act of God. So the Spirit of God moves in the womb of Mary, as the Spirit moved over the face of the waters of chaos at creation, to bring into being a new creation. Moreover, it might seem appropriate that Jesus, who was unique in his sinlessness and unity with the Father, should be unique in his birth, as he was to be in his death. It would certainly have given both Jesus and those who knew the secret of his birth a very clear testimony to his true uniqueness, should they have ever doubted it.

If Jesus was born while Mary was a virgin, that would certainly be a miracle. While it is just possible for women to give birth to females by parthenogenesis (without insemination by a male), women just do not have the Y chromosome that is necessary to produce a male child. If this happened, it is not possible within the limits of the laws of nature, so those laws must have been suspended. I have already suggested that, so far as science goes, God could suspend natural laws occasionally, for a good enough reason. Since the whole life of Jesus must be a miracle, in its exemption from sin, that seems a good enough reason for the birth of Jesus to be a miracle. This, after all, is claimed to be the decisive act of God on this planet to restore the divine purpose of creation and liberate humanity from its bondage to sin. So it is not just a rare event, but a totally unique one.

I am therefore inclined to accept that Jesus was, as the Apostles' Creed puts it, 'born of the Virgin Mary'. However, the symbolic value of the narratives can be retained even if its literal truth is denied. Since the virgin birth does not seem essential to the truth of the claim that Jesus is the Messiah of Israel and the incarnation of the cosmic Christ (as the resurrection perhaps is), I believe this article of the creed could be interpreted in a symbolic or metaphorical sense (to mean, 'born of a pure young woman who represents the "virgin Israel", the pure remnant of the people of the old covenant') without hypocrisy.

Making Present the Kingdom

If Jesus was aware of his unique relationship to the Father, was fully aware of the presence of the Father at all times, and was a perfect channel of the divine wisdom, power and love, then the kingdom, the rule of God, was already realised in him. The story of his life is the story of what happens when the rule of God meets the rule of human beings. When Jesus teaches that the kingdom is 'at hand', he means that it is actually present in him. When he calls his hearers to repent, to turn away from selfish greed, and prepare themselves for the kingdom, he is presenting them with the challenge to follow him.

Jesus is opposed above all to religious hypocrisy, to those who use religion to obtain status and wealth. Like the prophets of the eighth century before him, he warned of a great disaster to come on Israel if the people did not turn to God. Also like them, he promised the forgiveness and love of God for all who would open their lives to it. He taught that possessions and attachments should be renounced for the sake of the kingdom, for that relationship with God that would bring unsurpassable joy and unshakeable peace.

One of the extraordinary features of the recorded ministry of Jesus is that he did not concentrate on the pious and the good. He seemed to cultivate the friendship of the more disreputable, offering forgiveness to sinners and blessing to the poor and sorrowful. He renounced the idea of a political movement against Rome, and spoke of the kingdom as a non-violent movement of those who trusted solely in God and renounced anger, lust and possessiveness. Though he confined his mission almost entirely to Jews, he warned that the kingdom might be taken away from the nation of Israel and form a new community, bound by a new covenant with God.

Thus, he taught that divine wisdom is not the prudence that brings worldly success, but the commitment that renounces everything for the sake of the presence of God. The divine power is not the overthrowing of political oppressors, but the renewal of broken lives. The divine love is not a protection against all physical harm, but the promise of life that comes through the giving up of self.

In his teaching, Jesus is the perfect channel of the divine wisdom. For Christians, he is the heart of Torah embodied in a human life. In him the two great commands of Torah, to love God with all one's heart and mind and soul, and to love one's neighbour as oneself, find unrestricted expression. The divine wisdom may look foolish to those who put prudence first. But in a world broken by greed, hatred and ignorance, it is the true path to the realisation of God's purpose.

In his own life of renunciation of home and family, of healing and forgiving broken human lives and of accepting suffering and death rather than compromise with those in authority, Jesus sets the pattern for the life of the Church. The Gospel record of his life is not merely the story of a long-dead hero of faith. It is a pattern for the life of the community of the Spirit, as God seeks to re-establish the divine purpose for humanity in an estranged world.

In his actions, Jesus is the perfect channel of the divine power. He healed the sick, exorcised demons and, according to the Gospels, performed supernatural acts of great symbolic significance – turning water into wine, walking on water, multiplying loaves and fishes, and stilling storms. These acts are miraculous, in that they all transcend the normal powers of nature. They all have the symbolic function of manifesting the liberating, life-giving power of God over the forces of chaos and destruction that seem to control human existence. If one is disposed to think that in Jesus the divine power is perfectly manifested, that his very existence transcends the bounds of 'normal' human nature, then one would perhaps be surprised if there were no acts transcending normal powers and showing the power and purpose of God. If Jesus is what his followers claim him to be, the occurrence of miracles is not vastly improbable, but rather likely.

The miracles recorded in the Bible are usually performed in response to the prayers of the prophets, of people closely related to God and with a special calling in relation to the working out of the divine purpose. It seems that, at certain critical points in history, God's intentions and the religious sensitivity of some individuals come together in a way that releases supernatural power into the world, in extraordinary manifestations of the divine presence and purpose.

Christians will think that God is always responsive to human prayer and always inviting and challenging human response in a continuing dialogue of faith. This does not normally happen in miraculous ways, but by a continual shaping and reshaping of the events of life within the usual parameters of physical law. It is obscured and affected by human sin and ignorance, which often both corrupt the way the divine invitation is interpreted and impede the human response to it. So God's providential acts, which are directed to the ultimate welfare of human lives, are inextricably mixed with the obscuring factors of greed, hatred and ignorance.

God always responds to prayer, but those responses will take into account the preservation of the general intelligibility of nature, of general human freedom and of human insensitivity or opposition to the divine will. In general, God will seek to influence events without determining them unilaterally. What happens will depend partly on the receptivity or stubbornness of human hearts. It will also be affected by the fact that God knows what will best discipline human lives in virtue and in the love of God, so that the specific things we pray for may well well be more harmful than helpful to such goals.

In a deeply interconnected universe in which freedom, community and growth in divine–human relationship are primary values, prayer is not a magical mechanism for getting God to do things we want. It is the attempt to live consciously in God's presence, in awe, gratitude, penitence and loving devotion. Naturally, our concerns for others will form part of this relationship, and we will seek to channel divine power to help others by our requests to God, as well as by our actions. Because of this, we will always subordinate our desires to the will of God, but trust that our prayers will be used by God in influencing the future for good.

Those saints and prophets whose lives are closely united to God will be more receptive to the divine power, and God will be able to respond to them more obviously and effectively. It is not surprising that Jesus, united in the closest way possible to God, channels God's power to heal in a spectacular way. In his time, many forms of mental illness were put down to demon possession. In exorcising demons, he is dealing with dissociated parts of human personalities, bringing integration and healing to disturbed minds as well as to diseased bodies.

As with the virgin birth, the miracles of power over natural forces are chiefly important for their symbolic teaching, the way in which they express Jesus' role as Lord of the natural world, as the manifestation of God on earth, and as the renewer of the ancient covenant with Abraham and Moses. Some will take them as legends which present that symbolic teaching in objective physical form. But their physical occurrence may also be seen as an appropriate and miraculous manifestation of the kingdom breaking into the natural world, to show in an unrepeatable way that in Jesus the fullness of divine power is embodied on earth. The Gospel record provides an image of the divine nature, in a human life of reconciliation, healing, forgiveness, compassion and selflessness, and it provides an enduring exemplary pattern for the life of the Church. Its recital in the community of the Church becomes the mediator of the divine power that was in Jesus, the power by which the

Church also lives. For Jesus is also a perfect channel of the divine love, and in his life power is always in the service of love, a love that gives itself to the uttermost, so that others may find life.

The Final Sacrifice

As opposition to his teaching grew among the authorities, Jesus foresaw his own death and prepared himself to offer his life so that the kingdom, already present in his person, might come with power to his disciples. He gathered around him an inner group, the 'Twelve', who would form the core of the new Israel he foresaw. He promised that his death would not be the end, but that the Father would raise him up from death to become a mediator of the divine life for all future generations. The ancient prophecy that the Messianic kingdom would never end would be fulfilled, for when he was raised to his true glory and power, he would live with God for ever, to give the divine life to all who followed him.

In the mysterious rite of his final meal with the disciples before his death, he took bread and wine, saying that they were his body and blood, and that by receiving them the new covenant of the heart, prophesied by Jeremiah, would be sealed, and the Christ-life, the life that was seen in him, would be given to men and women who received it in faith, for as long as human life continued to exist. Bread is the staple food of life, and as the life of the Spirit filled the humanity of Jesus and shaped it into the form of divine love, so that life is given to the disciples of Jesus, to shape them into a form like his. Wine is the blood, the human life offered in the perfectly self-sacrificial prayer so that healing and reconciling divine power could be released into the world.

By sharing physically in that bread and wine, the disciples came to share spiritually in the life of Christ, so that it would live within them and be mediated through them to the world.

In the person of Jesus, God had united humanity to the divine life from which it had become estranged. Now, through sharing in the body and blood, the presence and life, of the Christ, the followers of Jesus could become 'partakers (*koinonoi*) of the divine nature' (2 Peter 1: 4). The new covenant is not the giving of a set of laws for a national state. It is the giving of the divine life to those who abandon themselves to the love of God. That gift is accomplished through the voluntary self-offering of Jesus, when he was mocked, scourged, crucified and condemned as a common criminal. Then, he experienced the desolation of abandonment by his followers, even the loss of the sense of God's presence that had suffused his whole life.

In offering himself wholly to the divine will, Jesus perfectly expressed the divine love, which shares in the suffering of all creatures. At that point in human history, human and divine love coincide. The cross of Jesus' death becomes the central symbol of the divine love that brought the universe to be. In the repetition of the last supper of the Lord, in obedience to Jesus' command, Jesus' perfect sacrifice is made present, the offering of prayer which is, uniquely, at the same time the self-giving of the life of God. Human sin is forgiven and human lives are united to the life of God through the cosmic Christ, who was manifest in Jesus and who continues to be present for all human history through the sacrament of bread and wine. The kingdom of God was made present in the life of Jesus, who was the historical incarnation of the eternal Christ. It is still made present wherever the bread and wine of the new covenant mediates the reality of Christ. In an admittedly incomplete, but nevertheless real sense, in the Eucharistic sacrament God continues the incarnation by forming those who share in Christ into the body of Christ on earth. In this way, it is through the fellowship of the Church that God presents, on this planet, the paradigm case of the restitution of the divinely intended purpose for the whole cosmos.

The Gospel of the Resurrection

Jesus' death at the hands of the religious and political authorities must have seemed to many the end of the Messianic dream, the final judgment of God upon a false prophet. The extraordinary fact is that it was to become the paradigm expression of the redemptive activity of God for millions of believers in subsequent world history. Jesus was crucified as a possible political agitator – although he never made any political claims – and as a religious heretic – although he taught total commitment to God, the rigorous upholding of Torah and the dawning of God's promised kingdom. It was his uncompromising challenge to all powers of human oppression and to all forms of religious hypocrisy that brought out the latent forces of intolerance and hatred that lurk within every human heart. He died as a martyr, a witness to the truth of God's uncompromising call to justice and universal love. But his death was more than that. It was a conscious and voluntary sacrifice of his life, in order that God's purpose for the earth might be made effective through him. He surrendered his humanity completely to God, so that the divine life could be mediated to an estranged world through him.

God honoured the self-sacrifice of Jesus, for his death on the cross, a death in shame and dishonour outside the walls of Jerusalem, the sacred city, was not the end. When Jesus sacrificed his life in order that the Church might come into being as the new community of the divine love, he did not simply die and pass into history. What the twelve apostles and a group of close disciples experienced, to their awe and astonishment, was the living presence of Jesus among them, after he had died and had been buried. Nothing else would have enabled them to recover from the shock of seeing God's purpose apparently defeated by the death of the Messiah and his rejection by Israel as a common criminal. Now that he was dead, how could the kingdom come, and how could Jesus' claim to be the anointed one of God be justified?

The resurrection of Jesus is the first of many divine acts that brought the puzzled apostles to believe that the kingdom was not what they had expected, the triumphalist expulsion of the Roman imperial power. It was not even the restoration of the twelve tribes to their ancient destiny. It was something new and strange, for which Jesus' life and teaching should have prepared them, and did prepare them, but not in a way that they had ever clearly recognised while he lived among them.

They were not expecting the resurrection – according to the Gospel accounts, Jesus had warned them that he would be killed and would be raised by the Father from death, but they had not understood what he meant. They thought the resurrection would be at the end of the world, when all the dead would rise to be judged. How, then, could Jesus be raised from death, and what did his unique resurrection mean?

The first disciples followed Jesus because he seemed to them to possess an authority and power that set him absolutely apart. He healed the physically and mentally ill, he called them to give up everything and follow him in trust, he promised the dawning of a new age when God would rule among his people in a new and living way. His death at the hands of the religious and political authorities came as a complete shock. Demoralised and bewildered, they gathered together in Jerusalem and tried to piece together in a new way the meaning of what had happened.

It was as they met that Jesus, whom they had seen crucified and buried, appeared among them, and it was over the next few weeks that the risen Lord brought them to understand at least something of the meaning of his life. They came to see that Jesus had never been a political leader who would restore the tribes of Israel, expel the Romans and reinstitute the monarchy. From the first, he had been the Suffering Servant, the bearer of the sorrows of humanity, the sacrificial lamb whose death would liberate, not just the Jews, but all humanity. He proclaimed a kingdom that was not a political state, not 'of this world', but a community of those who would turn from selfish desire and receive the eternal life of God as their only ruler.

What Jesus had proclaimed was that a new age of the Spirit was dawning, coming with power in his generation. He called his hearers to prepare to receive it with penitence and trust in him, renouncing all things, so that the Spirit could recreate them, forgive them, bear fruit in them, bring them to maturity and give them eternal life. Though the old covenant would never be overthrown, a new covenant of the Spirit was now to be offered to the whole world. In the end, it would transform the whole creation. His hearers could not know when that would be, but they would see its beginnings in their own lives.

Jesus had died, mocked as 'King of Jews'. But in his life he had always mediated the power of God, in his joy, wisdom, love and selfless compassion. That divine power had broken through in visions of his transfiguration, of his authority over nature, and in his healing power. At his death, God annihilated his human body and raised him to the divine presence in a transfigured body of light and glory, capable of fully expressing the divine, in so far as it can be expressed in human form. But before that final transfiguration of the human to be a perpetual manifestation of the divine Word, Jesus appeared to many in the world of the dead, and, over a period of six weeks, at various times to his disciples on earth, to teach them the true nature of the eternal kingdom of God.

15

The World to Come

The Hope for the World to Come

The Apostles' Creed states that Jesus, after he had been crucified, was buried and 'descended into hell'. Very little is said about this part of the mission of Jesus, which remains almost entirely hidden from us. The scientific worldview, too, has nothing to say about those who have died. The law of entropy decrees that, in the very long run, all physical energy in this universe will run out into darkness and inactivity. As far as physical science can see, this universe is finite, and is destined to come to an end. The planet earth will cease to be habitable much sooner, in about five thousand million years, as the sun expands on its own way to death. In any case, all life-forms die, and any meaning the process has must be contained within the short life spans of creatures on some of its inhabited planets (if indeed there is more than one).

That does not deprive the process of meaning, since a process can be perfectly meaningful even if it lasts only for a short time. Moreover, a Christian view of God suggests that God knows every creature perfectly, and that no good will ever be forgotten. All that ever has been will remain in the mind of God, cherished for what it was and for what, in the life of God, it always will be known to be. There is a certain profundity in the thought that we will be remembered by God for eternity, and an importance in hoping that we will be remembered for good and not for evil. For many Jews, that is a sufficient hope of immortality. Anything else, any desire for a continuation of individual life, might seem to be unduly egoistic and so self-defeating from a religious point of view, which tries to overcome egoism above all. So the important Jewish hope is not for personal immortality; it is to leave behind a good name, to be remembered for good by God.

Yet as the prophets reflected on their faith in the God who had called them into a covenant relationship, it began to seem that God had called them to something more than they naturally had a right to deserve. Had not God called them into an eternal relationship of unbreakable love? Had not God promised them the *shalom*, the fulfilment of responsibly created justice and peace that was the divine will for the universe? Such a fulfilment might lie far ahead in the future, but it seemed that the vast majority of souls would never enter into it. Where was God's love for the millions who died struggling, starving, giving their lives for the kingdom, but never seeing it? Would God's memory of the past not be largely a memory of pain and failure? And could God's kingdom ever be realised while humans remained free to fail again and again?

So among the prophets of the covenant of God there grew a hope for a future, new creation, in which all those who had died in failure and frustration could share in the love of God and delight in the divine presence. This was the hope of resurrection, of a world where the dead would rise to a new life in closer union with God and freedom from pain and death. In this resurrection world, the whole creation would be renewed and enter into its intended destiny. Thus, towards the end of the Hebrew Bible, the hope of the covenant people became hope not just for a future earthly society of peace, but for a new creation, a new heaven and earth, wherein the resurrected dead could share.

The biblical doctrine of the soul was consistent with such a belief in resurrection, for it saw the soul as properly embodied, and yet as not essentially confined to a specific human body in its present state. Might it not be possible for each body to be recreated in a more perfect form that could realise its inherent potentialities, so often frustrated by conflict and pain on earth? In such a world, persons could continue to grow for ever in creating, enjoying and sharing the endless goodness of divine creation. This world would have played its part in developing souls for the resurrection world, but it would be fulfilled beyond the confines of this space and time.

As the biblical hope for resurrection developed, it took many different forms. Some Jews, like the Sadducees, never accepted it. Some thought only the righteous would rise from death, while others thought that all the dead would rise. There were various views about what the resurrected body would be like. The crucial New Testament passage about resurrection is in the first Letter to the Corinthians, chapter 15. There Paul rejects both the view that the risen body will be exactly the same as this earthly body – 'flesh and blood cannot inherit the kingdom' – and the view that souls will exist in a disembodied state. Stressing that we cannot imagine what our bodies will then be like, he nevertheless states that they will be incorruptible, beautiful and imperishable, as different from our present earthly bodies as wheat is from the seeds from which it grows.

Paul's doctrine was based on his own experience of the risen Christ on the Damascus road – a body of light, capable of appearing in this world, but in reality existing far beyond the physical limits of this world. It gave rise to a remarkable new perception of the earth as a field where souls are prepared for eternity. Jesus' parables of the kingdom, in which seed is sown and grows until it is ready for the harvest, depict human souls as born into a world where good and evil, loving community and destructive egoism, are in desperate conflict. But in that world, God comes to draw souls to the divine by making them sharers in the divine self. As souls are united to God, they already begin to share in the resurrection life of Jesus. They become parts of the body of Christ, filled with the Spirit, bound indissolubly in love to the Father. They are transformed by the hope that their bodies of conflict and death will be transformed into bodies of the true communion of love and vehicles of vibrant life.

It was in the resurrected body that Jesus appeared to his disciples after his death, in a strange but wholly convincing manner. According to the Gospel accounts, he appeared to them in a locked room in Jerusalem. He walked, unrecognised, for seven miles, with two of them, before they suddenly knew him in the breaking of bread. Mary mistook him at first for a gardener. He was present, and then as suddenly was gone, at moments of his own choosing, and where he was, none could follow. Finally, his risen body disappeared from the world for ever, merging with the *Shekinah*, the cloud of the glory of God. This was no ordinary body of flesh and blood, walking out of the tomb and hiding somewhere in Jerusalem or Galilee. This was a resurrection body, transfigured by the beauty of the divine light, more real than mortal flesh, but as far beyond mortality as eternal life is from worldly death.

For the members of the new Christian community who had seen the risen Jesus, resurrection was no longer a postulate of faith that God's justice might require a new heaven and earth for its realisation. It was an experiential knowledge that the way of self-sacrificial love, which inevitably ends with the destruction of the earthly body, is in fact the doorway to a fuller form of transformed embodiment, in a world beyond this physical cosmos. Seeing the risen Jesus, they saw the life of resurrection to eternity, and they knew that the eternal kingdom of the divine love had touched this world and changed it for ever. In the resurrection of Jesus, the disciples saw the assumption of human life into the life of God. They saw the kingdom come with power, and the angels of God ascending and descending on the Son of Man, the one who opened up the path to eternal life for all humanity.

The hope for resurrection and for the final realisation of the kingdom was a hope for the whole earth. It was not, after all, a hope for the political primacy of the Jewish people. For that more universal hope to be realised, evil would have to be completely eliminated, and that could only be done when this form of spatio-temporal being had run its course. God became incarnate to assure humans that the divine intention was not destroyed in an estranged world, to unite souls to God and lead them to fulfilment in God. Yet God did not do that by exterminating evil, by an act of sheer power. Rather, by the force of love, Jesus called men and women to give up all and follow the way of love, in a world in which evil still remained.

That is why the Jewish people in general could not see Jesus as Messiah. He did not eliminate evil and imprison the torturers. He did not bring peace and justice to the world. How could he be the Messiah of God? The Christian answer requires a radical revision of the idea of the Messiah. It requires a radical revision of the idea of divine power and love. It requires one to accept that God acts only by the persuasion of love, to lead all who will respond to a kingdom that can only fully exist beyond the boundaries of this earth, in a new creation. The kingdom will come, but it will fully come only by love, and only beyond the barrier of death. In this life, it comes in weakness and ambiguity, but Jesus still calls many to enter it and to wait in hope for its consummation, when all evil will be eliminated from creation.

The Universal Hope for Salvation

If evil is to be eliminated, however, what will happen to those who are themselves evil, or trapped in evil, or unable to perceive the good? At this point, many Christians have fallen back into sub-Christian ways of thinking and looked forward to the torture and punishment of the wicked. Not realising that we are all only saved by divine love, not realising the extent and patience of divine love, not realising that in condemning evil we condemn ourselves, often adding hypocrisy to the list of our offences – not realising these things, Christians have supposed that God's justice requires the punishment of the wicked in everlasting hell.

Jesus shows that God will go to any lengths to reconcile the evil to the divine. Of course, those who have caused great harm must come to see what they have done. They must come to feel what such harm is like. They must be brought to feel and know the gravity of their deeds. That is a genuine punishment for those who are set on the way of self-will, and it is possible that egoistic souls will never come to feel contrition or sorrow for what they have done. The possibility of hell, of the torment of self-exclusion from love, exists. Jesus warns of its reality, in an attempt to turn his hearers to repentance and true worship of God, to learn the ways of love by seeing the consequences of the absence of love. But, surely, it is also possible that many may come to see the horror of evil and the deep attractiveness of the divine love, the possibility of forgiveness, and of making some amends for what they have done, by some form of self-giving service of others. If that is possible, a God of infinite love will never cease to hold out such possibilities, as long as there is anyone who may yet respond.

Can we accept that a God of infinite love will cut off every possibility of repentance at the moment of death? Or can we believe that such a God will only offer repentance and eternal life to those who have heard a particular version of the gospel of Jesus Christ? In seeking an answer to these questions, it is important to keep a firm grasp on the real fundamentals of Christian faith. These are that God is a God of limitlessly self-giving love, who 'desires all to be saved' (1 Timothy 2: 4). God unites the humanity of Jesus indissolubly to the divine Word, so that the power of egoistic desire over human nature should be broken. And God, through the Spirit, unites all who respond in faith to the divine life, so that 'as in Adam all die, so in Christ all will be made alive' (1 Corinthians 15: 22).

If God desires all to be saved, God will assuredly make it possible for all to be saved. We may be certain, then, that God offers eternal life to every rational soul, whether or not it has ever heard of Jesus. How God does this we may not know, but we may be sure that God speaks in some way, however hidden, to every human heart, and from each one God elicits a response that could lead to eternal life. All that is required of each human soul is that it honestly responds in commitment and trust to the call of love and compassion, however it is heard and in whatever form it comes. A God who limits salvation to the members of one religion, church or sect is unworthy of worship, and very different from the God revealed on the cross of Christ.

The God who saves is the God of unlimited love, truly revealed in Jesus Christ. All who are saved are, in fact, united in the cosmic reality of Christ, even if they have never heard of Christ. This means that many millions of souls are saved by a God they do not know, who has spoken to them in an unrecognised form. 'Many will say to me on that day, when did I feed you?' (Matthew 25). Any feeling of oddity this produces can be alleviated, perhaps, by distinguishing between being set on the path to salvation and to sharing in the divine love and being finally, consciously reconciled in the love of God. Millions of souls on earth can be set on the path to eternal life by their honest response to the claims and opportunities they discern in their lives. But the final realisation of what eternal life is can only come to those who recognise and worship the true Word – something even many Christians may find very different from what they have imagined.

In that sense, final salvation comes only after death, in the resurrection life. But there are many paths that will eventually lead to salvation – 'in my Father's house there are many resting-places' (John 14: 2) – for those who continue to trust in the highest insights they have. What this suggests is that there will be the possibility of progress, of learning and growing, after death, and before the resurrection kingdom is finally realised. Catholic and Orthodox Christians have long felt that few of us will be ready for the kingdom when we die. There will be much for us to learn, and much to unlearn, before we are ready to live unrestrictedly in the pure love of God. Those who have never known of Christ will learn what Christ is. Those who have known something of Christ will learn more of what Christ truly is. We will all learn more of the harm we have done, of selfish desires not yet overcome, and of the depths of love we have not yet begun to explore. The doctrine of purgatory is one way in which Christians have set out a view of a continued development in the love of God after death and before the final decisive elimination of all evil from the kingdom. That doctrine needs to be extended to embrace the full width of the redeeming love of God, so that even in hell the possibility of repentance is not closed and the hand of God is extended to all who will take it, without exception.

The 'Descent into Hell'

In Jewish tradition, even before the growth of belief in resurrection, the dead were often given a sort of shadowy half-existence in Sheol (in Latin, Hades, misleadingly translated as 'hell' in the Apostles' Creed). Sheol was not thought of as a desirable form of existence. Job 10: 4 describes it is 'a land of deep darkness, of deepening shadow, lit by no ray of light, dark upon dark'. It is the place where the spirits of the dead live or sleep, where 'there is neither doing nor thinking' (Ecclesiastes 9: 10). However, in the very varied set of beliefs in Hebrew thought there was also a belief in 'paradise', a much more pleasant abode of the dead, where there were shady trees, cool streams and pleasant fruits.

When Jesus speaks of life beyond death, he speaks in one parable of 'Abraham's bosom', of a great gulf fixed between those who suffer in Sheol and those who enjoy the blessings of paradise. He promises the penitent thief that he will be with him that very day 'in paradise'. His parables often depict a great separation between the 'wheat', gathered into barns, and the 'chaff', thrown away. They refer to a great feast, from which the impenitent are excluded, wandering in outer darkness. They speak of *Gehinnon*, the valley outside Jerusalem where rubbish was taken to be burned, where 'the worms are undying and the fire is not quenched', as a place where the corpses of the impenitent will be destroyed, or as a place where 'everyone will be salted with fire' (Mark 9: 49).

In the afterlife there is both judgment and blessing. There is Sheol, where the impenitent are punished by their exclusion from the love of God, by the burning flames of their desires, and by their personal experience of the harm they have caused to others. And there is paradise, where the dead are refreshed by the friendship of those who love God, but where they still await that final resurrection in which all the redeemed will share. Neither Sheol nor paradise should be thought of as the final resting places of the dead, for there yet lies ahead, at the end of historical time, the great Judgment and the new heaven and earth, the resurrection world.

In the afterlife worlds of Sheol and paradise, the dead do not yet have their glorious resurrection forms of embodiment. We may perhaps think of them, as the Hebrew Bible sometimes does, as *rephaim*, image-bodies, temporary forms for expressing their personalities, which allow for purification and learning. They are a little like the bodies we sometimes imagine in dreams, and so the afterlife worlds are sometimes spoken of as states of 'sleep'. In that sleep souls continue to be prepared for the Final Judgment, for the final division of good and evil that will bring the history of this universe and all the souls it has generated to an end.

It was those worlds of the dead into which Jesus descended after his crucifixion. There is only one New Testament passage that speaks of his mission there, 1 Peter 3: 19–20. The passage may be a legend or a piece of imaginative speculation, but, on the other hand, it may record an early tradition derived from Jesus himself. What it says is that after his death Jesus preached to the spirits in prison. Those spirits are further described as spirits who had been alive at the time of Noah. The text is admittedly unclear, but if those spirits are the souls of humans who had fallen under divine judgment at the Great Flood, then the implication is that Jesus preached to those who had rejected God during their earthly lives. I think this is the natural reading of the text. If so, it presents, in metaphorical fashion, a belief that repentance is possible after death, when even those who have rejected God come face to face with Christ as he really is, and are challenged to repent and believe the gospel.

Whatever the provenance of that particular story, I think that only such a belief in the possibility of repentance in Sheol is consistent with the unlimited desire of God that all should be saved. Otherwise, millions have died who have never had any sort of moral choice, or any realistic earthly chance of penitence, because their lives have been warped and corrupted by others. The traditional doctrine of purgatory has been reluctant to allow such a possibility of repentance. But if growth and purification is possible after death, it is a very small step to affirm that genuine repentance is also possible. Certainly many Jews and Muslims assert that hell is not permanent, and that God will finally deliver all from its pains. It would be ironic indeed if Christianity, the religion of the unlimited self-giving love of God, proved to be less charitable than its two sister Abrahamic faiths.

If so, we may say that Christ appears in the afterlife worlds, not only to encourage those on the way to salvation, but to call to repentance those who have rejected the ways of love on earth. Some may reply that in that case God's grace is cheap, and we can live how we want now, knowing we will have ample time for repentance later. Such a view does not cheapen grace, it cheapens any sense of human responsibility for our world. The reason for repentance is not to gain eternal life for ourselves, but to try to stop corrupting God's purpose in creation. The sufferings of Sheol, caused by our own actions, are real and grievous, even if they do not literally last for ever. But the most real suffering is the separation from God that we shall endure, our exclusion from the wedding banquet, from the community of divine love. Every rejection of God on earth makes that separation longer, and that will become a source of the deepest pain and regret in Sheol. It does not make our earthly actions less important to say that we will all be offered the chance of penitence in the world to come. For what really matters is whether or not we do God's will now. If we do not, we contribute to the frustration of God's purposes, and we actively help those who cause the suffering and death of Jesus. It is not cheap grace that goes on offering love and forgiveness even in the face of our petty destructiveness.

The Church as the Body of Christ

The resurrection of Jesus is an event of cosmic significance, because for those who can see it for what Christians believe it to be, it makes present on this planet the final goal towards which the whole creation moves. The final purpose of creation is the existence of a community of love and fulfilment in God, in which human lives are taken into the deepest form of union with the creator, and evil and self-will are destroyed – or more truly have come to destroy themselves. Perhaps the evolution of life on earth could have moved gradually but inexorably towards this goal, by the co-operation of humans with the Spirit of God, growing in their ability to manifest in history the archetypal form of the eternal Christ. But selfish desire turned the earth into a battleground of competing wills, and alienated the human world – what the New Testament calls the 'world of the flesh' – from the presence and power of God.

From within that alienated world, God took a human life, the life of Jesus, and united it to the divine life, so that it would be a true image both of what humanity should be and of what the divine love is. The death of Jesus, freely accepted by him, reveals and was always intended to reveal the self-giving love of God, which goes to any lengths 'that all should reach repentance' (2 Peter 2: 4). The resurrection of Jesus reveals that the divine purpose cannot be destroyed by self-will, but that it will now be realised, not on this planet but in the world to come, in a new creation that springs from this one but surpasses it in beauty and perfection.

The form its realisation will take is the assumption of every accepting human soul to be a sharer in the divine nature and a mediator of the divine love. In his resurrection appearances, Jesus shows this universal destiny, of rebirth in the world of the spirit, to be already accomplished in him, and to be prefigured at this point of space and time for the whole of this creation. Jesus makes present our future, and in him the goal, the end, of all things is already present. All that remains for us is to accept the Spirit that united his humanity to the divine, and allow the Spirit to shape in us the Christ life that will take us to share in that goal. In this sense, Jesus is the end of all things, and calls us, by sharing in his life, to strive towards the goal that is already existent in him.

The resurrection of Jesus is thus not just the reappearance of a human being after his death. It is the revelation of the final purpose of creation. When his mission on earth was completed, Jesus is said to have 'ascended into heaven, to sit at the right hand of God'. That is, he becomes for ever the human form of the formless God, the one who manifests and mediates the divine nature in a way that is suited for humans to understand. He is the human form of the eternal Word, who may have many forms, completely unknown to us, but who truly appears to us in the personal form of Jesus, the young Middle Eastern preacher and healer.

The Letter to the Hebrews speaks of the risen Jesus as the High Priest of the earth, continuing to offer the sacrifice of his own life, as a prayer that humanity might be reconciled to God (Hebrews 8: 1–7). His life of total self-surrender to the divine will is the pattern for all human lives. It is more than a pattern, it is the cause that begins to transform human lives into conformity with that pattern, and the goal towards which our lives move. The human form of Jesus already unites human nature completely with the divine, but that individual human life is only one way, if the definitive way in human history, in which the eternal Word is manifested.

The Word also becomes present on earth in the ritual activity of blessing and offering bread and wine in remembrance of Jesus. It may seem odd that bread, an inert, unconscious substance, should be a form of the local presence of God. It is the connection with the person of Jesus that makes such a thing intelligible. The life of Jesus was a form of the local presence of God, for in Jesus' self-surrender to the divine, the divine was able to act and to be known in and through him. It is not bread and wine as such that communicates the action and presence of God, but bread and wine as intrinsic parts of an act of prayer, an offering to God that is intended to make present at a particular time and place the self-offering of Jesus in the presence of the Father. It is in that sacrificial act, explicitly intended to make present the self-sacrifice of Jesus, that the bread and wine become 'the body', the physical medium of the presence of Christ, and 'the blood', the physical medium of the self-offered life of Christ.

God is known through the complex of memory, understanding, a long chain of historical causality, present prayer and hope for union with Christ, which centres on the offered bread and wine. In other words, it is the presentation of the bread and wine in an act of remembrance, sacrifice and hope that enables God to act and be known as a loving presence, as that same presence that was manifested in the life of Jesus, and especially on the cross. The Eucharistic sacrifice is a true form of the manifested presence of the eternal Word, which was in Jesus the Christ.

Protestant Christians have sometimes reacted against what they see as an unduly 'magical' view of the Eucharist, which may seem to give human priests power to make bread and wine turn into God. This is not because they value the 'Lord's Supper' less, but because they stress more the free and personal action of God, who comes to them in the form of Jesus and in the Word that is the proclamation of his universal Lordship. There is now greater understanding between the Orthodox, Catholic and Protestant parts of the divided Church that the offering of bread and wine was instituted by Jesus himself and is a way of making the presence of Christ manifest in the community of disciples. All agree that Christ is really and truly present in the community of believers as they gather together in prayer and thankfully commemorate the life and sacrifice of Jesus on the cross. A renewed stress on the communal and participatory context of the Eucharist, or Mass, and a renewed understanding of the way in which Christ can be sacramentally present among his people, are now making possible the overcoming of old divisions between Christians, many of which originated with long-dead social conflicts.

Part of such a renewed common understanding throughout the Christian churches is the recognition that the Word is not only objectively present in the Eucharistic act, as if this did not necessarily involve the believer at all. The chief purpose of that act is to incorporate human lives into the life of Christ. In taking and eating the bread and wine, the Word enters the depth of the human person, beginning to make real there the actions of divine love. Paul speaks of the mystery of Christ as the presence of Christ within the heart, the hope of glory to come. Through Eucharistic communion, the Spirit of Christ is enabled to act to transform individual lives into the likeness of the perfected humanity of Jesus.

The Church, united in the Eucharistic rite, is a community of people who seek to accept the action of the Spirit to transform them into the image of Christ. In so far as such transformation takes effect, they become another form of the body of Christ, channels of the Spirit of service and love, as Jesus was. To say that the Church is the body of Christ is to say that it also, as a fellowship, is a form of the local presence and activity of the eternal Word.

In the rite of baptism, individuals share in the self-giving of Christ, dying to the world of greed, hatred and ignorance, and are received into the community of the Spirit. Christians again disagree about whether baptism should be a matter of conscious commitment, made in adulthood, or a sign of the grace of God that is freely given, even to children, before any conscious response is made, but virtually all Christians agree that baptism is the proper means of reception into the Church community and the normal entrance into full Christian discipleship. Through sharing the body of Christ in the Eucharist, and through sharing in the presence of Christ in the verbal proclamation of his Lordship, believers are formed by the Spirit into the body of Christ, called and empowered to be active on earth in healing, service and reconciliation.

The Christian Church is not some sort of social or political organisation, granted authority and power to rule over the world. Nor is it a completely world-renouncing group of people chosen for salvation, while all the rest of the world goes to hell. For all its obvious failures and divisions, the Church is called to be a universal community that mediates the spirit of love, patterned on the form of Christ, reconciling the whole earth to the divine life. The Church exists, not to save its own members, but to renew the earth and reconcile it to God through the persuasive power of love. In pursuing this vocation, it needs to be keenly aware of its continual failure – the Church lives under the judgment of God – but also of the healing and renewing power of Christ, given to it by the pure love of God.

16

The End of All Things

The Church and the Future

As members of the body of Christ, Christians become participators in the life of Christ, sharers in the divine nature and mediators of the divine love. But what is the relation of the Church, the body of Christ, to the world? At times, it seems to be one of complete opposition: the first Letter of John says that those who love God must hate 'the world'. The disciples of Jesus are called out of the world, out of the structures of ambition, hatred and greed, to a life of self-giving love and surrender to the absolute goodness of God. They can expect only crucifixion by the world, and hope only for a kingdom beyond it, when it has been finally destroyed by its own passions.

Yet at other times it seems that God's purpose in creation cannot finally fail, that the Spirit must eventually transform lust into love, misunderstanding into compassion, and enable a society of justice and peace to come into being. As the universe was created 'in Christ', so finally 'all things will be reconciled in Christ' (Colossians 1: 20) and every knee shall bow before a final disclosure of God's invincible love.

Christians have always been torn between these two possibilities for the future of creation. But they have always agreed on one thing, that the Christ will appear in power and glory to usher in the fulfilment of the divine purpose for creation. The *parousia* of Christ, the clear and unmistakable presence of Christ at the end of historical time, is at the same time the full manifestation of the fellowship of those who have been made sharers in the loving nature of God, finally free from evil, pain and sorrow.

The law of entropy, which decrees that this physical universe will end in a long empty darkness, suggests that the *parousia* must be or usher in a new form of creation. One may reasonably see the material universe as gradually, over billions of years, unfolding its primordial potentiality for complete self-awareness and self-directedness. One may see that the intellectual understanding and responsible agency that seems to be the final actualisation of the potencies of matter is not confined to the forms of its material genesis and embodiment. One can thus see the possibility of a transformation of this cosmos, by means of the self-directing understanding it has generated, into new forms of embodiment, expressing the life of the Spirit more fully and creatively than is possible in a universe doomed to eventual decay. Seen thus, this space–time universe is one phase in divine creation. It has irreducible importance, as the generating ground of souls. But the fulfilment of the lives it generates will be found in further phases of created being – phases we can hardly begin to imagine.

Paul's metaphor of the seed giving rise to the corn comes to seem very apt. This cosmos bears the seeds of conscious understanding and agency, but those seeds will bear fruit only when they leave the earth and grow into the light of the sun. Christ, the eternal Word, is that sun, drawing souls born into time towards the uncreated light. The Spirit, which shaped all things in the image of the archetypal Word, and which filled the life of Jesus as the exemplar of the divine love, continues to act throughout the cosmos to unite all things in the pleromal Word, the completed actualisation of the positive potentialites of created being. As Christ is the archetype of all finite forms of being, so Christ is the pleroma, the full perfection of created being. In Christ all beings, freed from suffering, will express and mediate the divine life in the forms of their own individuality. In Christ they will share and co-operate in one communion and harmony of being. As all things were formed in

Christ, so all things will find their completion in Christ. In that one integrated communion of being, all created things will be at last united and perfected in God.

The Church, as the body of Christ, empowered with the life of Christ, sustained by the presence of Christ in the Eucharistic sacrifice, is meant to be a temporal image of the Christ who is the archetype and pleroma of created being. It is, however, a broken, divided and ambiguous expression of the life of Christ. For it continues, and must continue, to invite sinners into its fellowship, and it claims only to be the fellowship of the forgiven, not of the perfect who need no forgiveness. The Church is therefore primarily a community of hope, not of attainment.

Its central rite, the Eucharist, not only looks back to the temporal moment of Jesus' self-offering; it looks forward to the marriage banquet, when bride and groom, the beloved people and the divine lover, become truly one, and division and ambiguity are put away. At that point all things in heaven and earth will be united in Christ (Ephesians 1: 10), and the renewed cosmos will itself be the body of Christ, the true expression of the life of Christ and the adequate means for enacting the divine purpose. Just as the Eucharist makes the past sacrifice of Jesus present and effective, so it makes the future consummation of all creation in Christ present and effective. It brings the promise of the completed future into the present, to inspire hope and confidence in the power and purpose of the love of God. It makes present the assurance of the coming of the kingdom with power.

For this to occur, all that is in conflict with good must be excluded from this renewed creation. So the coming of Christ in glory will be a taking out from this universe of all who are sharers in the eternal life of God, so that they can live in a new, incorruptible creation. The New Testament speaks of Christ coming on the clouds – the *Shekinah* of the presence of God – to take his own to be with him for ever. This should not be interpreted to mean that the physical body of Jesus floats over the planet earth and pulls believers up into the sky – a crude and unbelievable scenario. What is pictured is the manifestation of the eternal Christ, in a form far beyond anything we can imagine, and certainly beyond any human form, bringing the history of the cosmos to an end and inaugurating the new resurrection world in which all those who share in the divine nature, of whatever species or form of life, will exist.

Sacred Cosmology: the Apocalyptic Narratives

The two creation narratives of the Book of Genesis provide a sacred cosmology, expressing spiritual truths about the general relation of creator and creation in the form of narratives about the beginning of this space–time order. As one might expect, there is also a sacred cosmology that expresses spiritual truths about the final realisation of God's purposes in creation, in the form of narratives about the end of this space–time order, 'the end of the world'. The symbols for these narratives are taken from a particular style of apocalyptic literature ('apocalypse' simply means 'revelation') that was current before and at the time of Jesus. Many of the interpretations of these symbols have a localised cultural significance that is now unclear to us, but others refer back to symbolism used by the major prophets of the Hebrew Bible.

There are so-called 'apocalyptic discourses' in the three Synoptic Gospels (Matthew 24, Mark 13, Luke 21); some of the New Testament letters (particularly those to the Thessalonians) and the Book of Revelation provide parts of a general narrative of the end of the world. By decoding part of the discourse in Mark's Gospel the general character of these narratives can be discerned, and the elements of the underlying spiritual teaching can be drawn out and distinguished from the literal interpretation.

Mark's account, the 'little apocalypse', introduces the subject with a prophecy by Jesus that the Temple in Jerusalem would be destroyed. That did happen in 70 CE. It was part of the teaching of Jesus that his message posed a point of critical choice for Israel. Through him God offered the possibility of repentance and a new life for Israel, but he warned that the rejection of God's message would result in the destruction of Israel as a nation state. The destruction of the Temple, which made the central sacrificial rituals impossible and thus ended the institution of priesthood in Israel, was a terrible blow to Hebrew religion. It did not end God's covenant with Israel, but that covenant now had to continue in a quite different, non-priestly, way in the universal diaspora of Judaism. With Jesus, Christians believe that God made a new covenant, open to all who follow Christ, and instituted a new community, the Church, which was to be a sign of the kingdom of God in the world. This was not, of course, the end of the world, much less the end of the universe. But it was in a real sense a new age, a new era in world history.

When the disciples ask Jesus when this will be, Jesus warns that there will be many false Messianic claimants, and goes on to speak of wars, earthquakes and famines, which will be, he says, 'the beginning of the birth-pangs' (Mark 13: 8). This prediction of the 'Messianic woes' of war, earthquake and famine builds on the prophecies of Isaiah (Isaiah 19: 2), which refer to the defeat of the tyranny of Egypt and a future political and religious ascendancy for Israel. The Lord, says Isaiah, will 'smite and heal' the Egyptians (Isaiah 19: 22), and they will turn to the Lord, offering sacrifices in his name. So God's judgment on the political powers of the earth is a preliminary to their healing, and to a new age of justice and peace.

Judgment is a major theme of apocalyptic writing, which takes up and intensifies themes in Ezekiel and Daniel (Ezekiel 7: 2: 'An end has come upon the four corners of the land'). 'The end' is not the end of the world, but the end of Israel's age of apostasy and idolatry, and the end will be bloodshed and disaster for the oppressive nations of the earth. But judgment is never the final word. On the contrary, it always functions as a warning, and is never without hope for those who turn in penitence to God. 'The Lord will have compassion on Jacob and will again choose Israel . . . and aliens will join them and will cleave to the house of Jacob'(Isaiah 14: 1). The emphasis is not on the ending of the whole universe, but on the ending of a corrupt age and the beginning of a new age, when 'the ransomed of the Lord shall return and come to Zion with singing' (Isaiah 35: 10).

What the apocalyptic writings are doing is taking a familiar prophetic theme of the catastrophic downfall of a world set against the justice and loving-kindness that God requires. This theme is balanced, as it is in the major prophets, by a call to turn to God and a promise that those who do will be healed and united to God. They will be 'saved' (Mark 13: 12). These themes are referred to the particular case of the fall of Jerusalem, and the creation of a new 'chosen community' (*eklektous*), the Church, which would usher in a new age. But the themes are also presented in cosmic imagery drawn directly from the prophets, to stress their crucial importance in the working-out of the divine purpose on earth. On the Day of the Lord 'the sun will be dark at its rising and the moon will not shed its light' (Isaiah 13: 10). On that day 'the host of heaven shall rot away and the skies roll up like a scroll' (Isaiah 34: 4). In the passages of Isaiah from which these phrases are taken, Isaiah is talking about the fall of Babylon and of Edom, respectively. These historical events are used as symbols of the destruction of all the powers of pride and greed, hatred and ignorance, that oppose the loving purposes of God.

The central figure in Christian apocalyptic is the 'Son of Man', the title so often used of Jesus in the Synoptic Gospels, who comes 'with the clouds of heaven' to have 'an everlasting dominion' (Daniel 7: 13 and 14). His rule follows that of the beasts who have dominated world history, and it brings an age in which 'the saints of the Most High' will rule in justice. At that time, a 'great trumpet will be blown' (Isaiah 27: 13) and all the scattered people of the tribes of Israel will be gathered from the 'four winds' (Zechariah 2: 6) – 'if your outcasts are in the uttermost parts of heaven, from there the Lord your God will gather you' (Deuteronomy 30: 4). God will 'circumcise their hearts', so that they will love the Lord their God with all their hearts.

It is important, if one is not to fall into a naive literalism, to read the apocalypse in the context of prophetic thought, and to trace its imagery back to the primal Jewish concern with judgment on 'the nations', the restoration of Israel, and the eventual fulfilment of the divine purpose for humanity. In the Christian case, as has been noted, the Messianic idea was changed so that it was no longer concerned with establishing a national state of Israel. It is therefore to be expected that the 'coming of the Son of Man' would not be identified with the restoration of Israel as a state. It is concerned with judgment on evil, with a call to repentance, with the promise of a new age and a new covenant with God, and with the eventual fulfilment of God's purpose. But how exactly these things are to be understood is left to the sensitivity and understanding of the hearer.

The Marcan apocalypse refers to an imminent social catastrophe, allied with a new hope for the future. The catastrophe was the collapse of the civilisation of the ancient world, and an ensuing age of barbarism and anarchy. The military power of the Roman Empire would collapse, Jerusalem, the Temple and the sacrificial cult would be destroyed, and the state of Israel would cease to exist. The hope was the existence of a community of the new covenant, the Church of the new age, as a growing company of those gathered to God through Christ (gathered by angels from the corners of the earth to the presence of the Son of Man, clothed with the 'clouds', the *Shekinah* of God), and enduring to the birth of a new era.

These temporal symbols also look beyond historical time altogether. The whole universe is to be so transfigured that evil and suffering are eliminated, and a new creation will come into being, with Christ as its centre. Such transfiguration cannot occur until the last freely choosing human being has been born, and until the present laws of nature, which entail suffering to some degree, are changed. In other words, one is looking beyond historical time to a different realm of being. What is being said is that Christ, the cosmic Lord, rules in a kingdom beyond suffering and self-regard, and all who are faithful to him will find their lives fulfilled by their completed and conscious union with him.

Finally, the symbols are internalised to enable the believer to view each moment of life as one of judgment on evil and election to eternal life in Christ. The judgment is not on other people sometime in the near future. It is on oneself, now. The rapture and union with Christ is not in historical time at all. It is a process begun now, as Christ comes at each moment in judgment ('like a thief in the night': Matthew 24: 43), but also, and primarily, in forgiving love. The completion of the process is beyond history, and on a different plane of spiritual existence.

What has confused many Christians about the apocalypse is the failure to understand how the prophets use symbolic language to speak of the eternal realities of judgment, forgiveness and reconciliation with God. A naively literal interpretation has sometimes been given, even though such an interpretation should have been ruled out by the plain statement that 'This generation will not pass away before all these things take place' (Mark 13: 30). Only an interpretation of the general kind that I have provided is compatible with that statement.

There is a sacred cosmology of the 'end of the world'. But, like the Genesis narratives, it offers spiritual, not literal, truths about events that are depicted in symbolic images. There is a literal truth at stake. God's purposes for the universe will be fulfilled. Evil and suffering will finally be destroyed. All who accept God's grace will share in an eternal creativity, wisdom, appreciation and joy. That fulfilment will be in Christ, the cosmic liberator and ruler, who was incarnate on this planet in the person of Jesus. It is wise, however, not to press too far for literal truths about things we can scarcely envisage. As the first Letter of John says, 'It does not yet appear what we shall be, but we know that when he appears we shall be like him' (1 John 3: 2). The form of the glorified Jesus is beyond our imagination, and we do not know what we shall be like in such a transfigured existence. All Christians can say is that Christ lives in us, and he is 'the hope of glory' (Colossians 1: 27), of a transformation of our present individualities into forms of beauty and powerful love.

Four Scenarios of the Future

From the perspective of cosmic evolution, it is very hard to say what the future of the human race will be. I think there are four main scenarios one might envisage. One possibility is that we will destroy ourselves or be destroyed in the near future. We have the capacity to do so, and there are many natural catastrophes, like the impact of a comet, that could effectively wipe out human life. If that happens, a Christian view would be that God will create a new form of existence, a resurrection world, in which all who have lived on earth will enter into a fuller form of existence. The dead will in afterlife worlds complete whatever opportunities of growth or learning are possible for them, or they will experience the harm they caused to others in a purgatorial existence. Then will come the Final Judgment, when all the dead will leave the intermediate worlds of Sheol and paradise. They will rise to find themselves in the presence of the glorified Jesus, the transfigured human form who is the vehicle on earth of the cosmic liberator and ruler, the Christ. Jesus Christ will be the one by whom all human lives will be measured. His *parousia* will be the final incorporation of all things into the pleromal form of the Christ, when all the redeemed will form the body of God, being vehicles and mediators of the divine presence. We cannot transcend the symbolism to say just what literal truths are here expressed, but the Christian's hope is that part of what is symbolised is true and final human fulfilment.

Another possibility for the human future is that we might devise some form of genetic engineering that will enable humans to evolve into higher forms of life. At present, the dangers of genetic engineering seem to be unacceptably great, but in thousands or millions of years it is possible that more precise control will be possible. My own feeling is that human selfishness will not be eliminated by genetic engineering as long as humans remain free and morally responsible, so I am sceptical about the possibilities of breeding a perfectly just society. Maybe one cannot rule out the development of such a society in the very long term, whether by genetic control or by physical and social evolution. So it is possible, as Teilhard de Chardin thought, that humanity will evolve into a higher form of existence – what he called a 'noosphere' – and eventually realise the kingdom somewhere in this universe, if not on this planet. In that case, one may speak of the 'return of Christ' as the realisation of the unity of all things in the universe, or at least on this planet, within the realised pleromal form of Christ.

Even if that is so, however, this universe will eventually come to an end, so human existence will sooner or later transcend this space–time altogether. Christians will hope that all the dead will share in the realised kingdom of God, so if living and dead are to share the kingdom, it will again have to be beyond the bounds of our physical universe. Though this second scenario may seem more optimistic about the future of humanity in this universe, the ultimate human goal is, as with the first scenario, placed beyond this space–time.

A third scenario is that humans will be superseded by machines, by constructed artefacts with artificial intelligence and true consciousness. That would be equivalent to the dying out of the human species, and its replacement by other forms of consciousness. I can think of no reason why artificially constructed personal beings should not exist. If they do, they will have as much reason to hope for immortality, and for knowledge and love of God, as naturally reproduced organic beings have. There will be a Last Judgment for computerised intelligences as well as for humans, and they will take their place in the resurrection world with whatever other sorts of finite creatures there may be.

In this scenario, humans may continue to exist, no longer the most developed species on this planet, but still capable of creativity, understanding, and knowing and loving God. Eventually, like all life-forms, humans will die out, and their own proper form of fulfilled personal life will be completed in the trans-historical kingdom of God. There, they will almost certainly not be the most advanced form of creaturely life, but they will nevertheless possess immeasurable dignity and value, as beings touched by and included in the infinite life of God.

A fourth scenario, and the last I shall consider, is that humans may make contact with alien beings from some other star system. The interesting question here is what difference that would make to human religious beliefs. On this planet, Christians have usually assumed that it is appropriate to preach the good news of Jesus Christ to every human being. That good news is that the creator shares creaturely experiences and desires all responsible, conscious beings to become members of a co-operative community of creative love and knowledge that will be fully realised outside this space–time. If humans turn from selfish desire and receive the Holy Spirit, they will be united to God for ever.

Those beliefs are as relevant to extraterrestrial beings as they are to humans. I would expect that advanced alien cultures would have beliefs very similar to those beliefs. That is, a universally loving God would have revealed the divine purpose to them, and provided a way of realising that purpose. The specific human person of Jesus, however, is the human embodiment of the cosmic Christ, and I would not expect that aliens would need to believe that Jesus is their saviour. It is more likely that they would have their own embodiments of the universal liberator who exemplifies God's desire to unite all creatures to the divine being.

This does not mean that the person of Jesus would become irrelevant in a meeting of different life-forms. It would always be important that the creator had taken human form on this planet, in order that humans could become sharers in the divine nature. Jesus is not ever going to become an almost forgotten figure from some ancient and primitive human culture. Christians believe that he is the paradigm exemplar for the earth of divine self-giving love and of the uniting of humanity to the divine life. He is thus the earthly embodiment of the cosmic archetype upon which the whole universe is patterned by the creative activity of the Spirit. The same Spirit acts throughout the world to make the exemplary form of that archetype present in all human cultures. Beyond the confines of this planet, the Spirit acts throughout the whole cosmos 'to unite all things in Christ, things in heaven and things on earth' (Ephesians 1: 18). Christ is not only the pattern of creation, but also the 'liberator', the one who frees personal creatures from selfish desire, and unites them in one organic community, empowered and perfected by the Spirit. That final unity, which may only exist fully beyond this cosmos, is the pleromal Christ, the fullness of the divine embodiment in all creatures who have been liberated from self. In the pleromal Christ, all personal beings, of whatever life-form or galaxy, will be sharers.

The 'body of Christ' will be much more diverse and expansive than anything envisaged in the traditional, rather anthropomorphic, iconography of Christianity. Yet traditional Christians have always believed that countless numbers of angels exist, so they have never thought that humans are the only created beings. The human person of Jesus will always have cosmic significance, as the mediating form of God for one small planet. Christians can say that 'At the name of Jesus every knee should bow, in heaven and on earth and under the earth' (Philippians 2: 10). But they may find themselves bowing to many other finite forms of the cosmic Christ, all mediating for their own worlds the infinite reality of the divine Word.

Since we have no knowledge of the future, we cannot tell which of these scenarios is nearer the truth, or whether there are any other extraterrestrial responsible agents at all. Nevertheless, Christians can plausibly say that however many created beings there are, and whatever the future of the cosmos holds, nothing 'in all creation will be able to separate us from the love of God', that has been definitively shown in Christ Jesus (Romans 8: 39). The final future of this cosmos is not a blowing out into endless nothingness, but the beginning of endlessly new forms of communion with the creator.

Judgment and Salvation

One of the most powerful images of Christian iconography is the Last Judgment. Michelangelo's painting in the Sistine Chapel in Rome is one classical depiction of the division of humanity into saved and damned, with the saved being taken into the realm of light, while the damned, including at least one cardinal whom Michelangelo particularly disliked, are pitchforked downwards into realms of darkness and torment. The Last Judgment is the divine response to the whole history of this cosmos, and it may seem to imply that the eternal life of joy that the gospel promises is reserved only for a favoured few. The idea of judgment implies a division, and indeed the New Testament does teach that God 'will render to every man according to his works' (Romans 2: 6). Those who have done well will be received into the presence of God, and those who have done ill be be cast away, into the fire of Gehenna.

There is very good reason to doubt, however, whether this is an adequate Christian view of the final destiny of humanity. As Paul says, 'no human being will be justified . . . by works of the law' (Romans 3: 20). One has an uncomfortable feeling that on Judgment Day we will almost all be weighed in the balance and found wanting. The deeper Christian doctrine is that 'we are justified by faith' (Romans 5: 1), and that 'in Christ God was reconciling the world to himself, not counting their trespasses against them' (2 Corinthians 5: 18).

The attempt to do works of charity is not useless, but it is not sufficient to prepare us for full loving relationship with God. Perhaps, in the intermediate state, it may be that we are saved, 'but only as through fire' (1 Corinthians 6: 3). We virtually all have to learn a deeper penitence, a deeper self-surrender, and a deeper love. But in the end, the judgment of works has only one aim, which is to lead all souls to penitence, to acceptance of the love of God that alone can unite us for ever to divinity.

In this respect, it is impossible to put limits on the mercy of God: 'The Lord is . . . forbearing . . . not wishing that any should perish, but that all should reach repentance' (2 Peter 3: 4). Not only is it God's wish that all should be saved, God acts to realise that wish: 'The grace of God has appeared for the salvation of all men' (Titus 2: 11). According to the laws of retributive justice, all stand condemned, excluded from God's love. But according to the law of that divine love itself, God wills to unite all souls to divinity: 'One man's act of righteousness leads to acquittal and life for all men' (Romans 5: 18), and 'God has consigned all men to disobedience, that he may have mercy upon all' (Romans 11: 32).

The gospel message seems quite clear, then, that God desires all souls to come to repentance, and if they do, God will have mercy on them. The only thing that can prevent universal salvation, the union of all souls who have ever lived in the unending love of God, is a wilful and obstinate refusal to repent, in clear view of the eternal life that one is rejecting. We cannot know that all will come to repentance. But we can know that all will be offered sufficient opportunity to do so, and that God will not simply cut off the possibility of further repentance at some arbitrary point.

At the Last Judgment, what is revealed is that all are condemned by strict justice, but that the one who is the Judge has in person borne the penalty of judgment, so that the only word pronounced on that day is the word of mercy. If any stand finally impenitent, they must be finally removed from creation, in 'the second death' that awaits those who refuse love to the end. But all, however egoistic and destructive they have been, who bow in repentance before God, will be at last taken into the love of God and transformed by the community of the Spirit of God, which will then exist without impediment and in the fullness God has willed from the beginning of creation.

Conclusion

There are two great pictures of the nature of the cosmos confronting us in modern times. One depicts it as a purely physical construction out of blind elements, accidentally forming complex forms of life, but without purpose or objective value. The other depicts it as the creation of a spiritual being of supreme value and power, who is the only self-existent reality. For the former view, any talk of a purpose in life or of a form of existence beyond the physical is senseless. For the latter view, however, the universe exists to express the purposes of the creator. All that ever comes to be in creation is experienced and remembered by God for ever. And, because it is meant to realise a divinely devised purpose, it finds its fulfilment in God, as all that is good in creation is conserved in the being of God for ever. God sets out in creation to realise specific purposes. God experiences that realisation. And God conserves all created goods in the divine consciousness. Thus, all good is conserved and fulfilled in God.

If that is the ultimate nature of things, it becomes only natural to hope that the fulfilment of all created goods might be experienced, not only by God, but by all the finite souls that have contributed towards it, but have suffered so much frustration and pain in their own lives. If there is a creator God, that God could certainly bring created souls to share in consciousness of the fulfilment of created goods.

It might even be that God's own nature, as love, is only fully realised by the creation of other conscious agents with whom

God can share in fellowship, by giving, sharing and receiving a love that binds creator and creatures together in a community of spiritual being. If that is so, it is again natural to hope that such a community might make it possible for every created member of it to share in knowledge of its final fulfilment in God. In other words, the love of God might require that the fulfilment of creation is not only experienced by the one consciousness of God, but shared in a communion of love that God brings to completion.

In this way, the existence of a resurrection world, however exactly it is envisaged, comes to seem a natural hope for a created cosmos. The revelation of the existence of that world in the life, death and resurrection of a particular human being in a carefully prepared historical context would then be a natural expression of the divine will to bring souls into loving relation with God. The resurrection of Jesus, which seems in a materialistic context so irrational and puzzling, then comes to seem a natural and appropriate expression of the divine will to bring creation to its intended consummation. In that sense, the resurrection of Jesus and the hope for a universal resurrection to eternal life are natural outworkings of belief in a creator.

The Christian vision is that the whole cosmos is patterned on the archetypal Word, the unitary array of eternal forms of thought in the divine being, and brought into existence by the creative activity of the cosmic Spirit. The material cosmos is a vast emergent, interconnected and intelligible whole, which generates from itself forms of self-aware and self-directive being. The creator does not remain aloof and apart from this cosmos. Through the Spirit, God is a constant presence and co-operative agent in the process of emergent evolution.

When rational and responsible agents come to exist, God wills to unite them to the divine being, so that they can be vehicles of the divine purpose, intimately conscious of their divine ground and goal, fellow-workers with God in creating and delighting in new forms of goodness and community. On this planet, God acts in many ways to begin to realise that final purpose. In the life of Jesus, and in the complex of unique historical events in which that life is embedded, Christians believe that God has acted in a particular and distinctive way to reveal the divine nature as self-giving, redemptive and unitive love. That love becomes a transforming power in the lives of humans, as they are united by the power of the Spirit to become the 'body of Christ', the servant community of love, on this planet. Though human selfish desire and greed impedes the purposes of God, there are many communities in which such selfishness can be at least partially overcome, and union with a reality of bliss, compassion and wisdom can be achieved. The Christian Church understands itself as a community that is meant to be a paradigmatic icon of the kingdom of God, the trans-historical community in which God's purpose for this universe will be finally realised. It is such an icon, in so far as it expresses loving dependence on the Father, is patterned on and united in Christ, and is filled with and shaped by the Spirit. Within it, human lives can begin to share in the divine nature, and thus foreshadow the destiny that awaits every created personal life.

After earthly life, according to Christian belief, all personal agents are given existence in worlds of purgation and sanctification (Sheol and paradise), until they are prepared for their final fully informed decision to accept or reject the love of God (the Last Judgment). Then, in a new cosmos, all evil will be eliminated, and the Spirit will unite all things within the pleromal form of the Christ, the perfected form of material creation, when the renewed universe will be the true temporal image and sacrament of eternity. At that point, the divine purpose for this universe will be fulfilled, and created persons will share together in that infinite journey into the knowledge and love of God which is eternal life.

The scientific worldview enables one to see this Christian vision in its full cosmic scope. Increased understanding of the diverse worldviews and religions of the planet helps one to discern the distinctive witness of Christian faith within a fully appreciated and assimilated global context. A greater understanding of the varied and changing ways in which historical contexts influence religious imagery, and a clearer grasp of the symbolic nature of that imagery, encourages one to distinguish between the spiritual content of the sacred cosmology of Christianity and the literal forms in which that content has been expressed.

Christian belief in the new millennium will, as it always has done, take many different forms. Evangelical, Orthodox and Catholic forms of Christianity are increasing remarkably in strength in many parts of the world. It is in my view of major importance that they should take full account of modern scientific understanding, a global vision of faith, a sensitivity to the nature of religious symbolism, and an awareness of the varied cultural and historical contexts of religious belief. The interpretation I have given seeks to be true to the mainstream tradition of Christianity, in the light of more recent developments in factual and moral understanding. It thus presents one way in which I think Christian belief can continue to be a positive and creative factor in shaping life on this planet in the third millennium.

Selected Further Reading

This is a list of just ten main books, five on religion and science and five on theology, the contents of which underlie various parts of my text, and which themselves contain important bibliographies for further reading.

Ian Barbour, *Religion in an Age of Science* (San Francisco: Harper, 1990).

John Hedley Brooke, *Science and Religion* (New York: Cambridge University Press, 1991).

Paul Davies, *The Mind of God* (New York: Simon & Schuster, 1992).

Arthur Peacocke, *Theology for a Scientific Age* (London: SCM Press, 1993).

Mark Richardson and Wesley Wildman (eds.), *Religion and Science* (New York: Routledge, 1996).

Catherine Lacugna, *God For Us* (New York: Harper, 1991).

John Macquarrie, *Principles of Christian Theology* (London: SCM Press, 1966).

Gerald O'Collins, *Christology* (New York: Oxford University Press, 1995).

Karl Rahner, *Foundations of Christian Faith* (New York: Crossroad, 1995).

Richard Swinburne, *The Existence of God* (Oxford: Clarendon Press, 1979).

My own views are given at greater length in a four-volume work of comparative theology published by Oxford University Press, dating from 1994. The titles are: *Religion and Revelation*, *Religion and Creation*, *Religion and Human Nature* and *Religion and Society*.

Index